AF779890

PHOTOMICROGRAPHY DESIGNS AND PATTERNS

LEWIS R. WOLBERG

DOVER PUBLICATIONS, INC. NEW YORK

Copyright © 1974 by Lewis R. Wolberg.
All rights reserved under Pan American and International Copyright Conventions.

Published in Canada by General Publishing Company, Ltd., 30 Lesmill Road, Don Mills, Toronto, Ontario.
Published in the United Kingdom by Constable and Company, Ltd., 10 Orange Street, London WC2H 7EG.

Photomicrography Designs and Patterns is a new work, first published by Dover Publications, Inc., in 1974 under the title *Art Forms from Photomicrography*.

DOVER *Pictorial Archive* SERIES

Photomicrography Designs and Patterns belongs to the Dover Pictorial Archive Series. Up to ten illustrations from this book may be reproduced on any one project or in any single publication free and without special permission. Wherever possible please include a credit line indicating the title of this book, author and publisher. Please address the publisher for permission to make more extensive use of illustrations in this book than that authorized above.
The republication of this book in whole is prohibited.

International Standard Book Number: 0-486-20298-4
Library of Congress Catalog Card Number: 74-76837

Manufactured in the United States of America
Dover Publications, Inc.
180 Varick Street
New York, N.Y. 10014

INTRODUCTION

> "The highest demand that is made on an artist is this: that he be true to nature, study her, imitate her, and produce something that resembles her phenomena. When the artist takes any object of nature, the object no longer belongs to nature; indeed, we can say that the artist creates the object in that moment, by extracting from it all that is significant, characteristic, interesting, or rather by putting into it a higher value." Goethe (Introduction to the *Propyläen,* 1798).

Nature has always been a source of artistic inspiration to man. Sea, sky, and moon, sweeps of forest, snow-decked hills, and all living things have always excited poets, painters, sculptors and designers. It is little wonder that, as man's vision has widened through the miracle of optics, new perspectives on natural forms have come into prominence. Among these are the glimpses given us by the microscope into the construction of matter. An understanding of how the microscope works and the processes of photographing magnified images (photomicrography) will enable the reader better to understand the illustrations in this book.

The principle behind the microscope is a simple one. The closer the eye is brought to an object, the larger the object appears to us and the more detail we see. There is a limit, however, to how close we can come to an object with the naked eye. The lens of the eye will not accommodate itself to an object closer than 10 inches, and the image will be thrown out of focus. We can get closer to the area under observation if we place a simple magnifying lens in front of the eye—a double convex lens with a short focal length, for example. This method of magnification generally produces images that are not more than 10 to 15 times as large as the actual object. What greatly increases magnification is the use of several lenses superimposed one on another.

The simplest magnification system of this type is composed of two lenses mounted on the opposite ends of a tube, the one near the object being called the objective, the one near the eye, the eyepiece. This is the principle of the compound microscope. The total magnification factor achieved is the product of the magnification of the objective and the magnification of the eyepiece.

The first compound microscope was constructed c. 1590 by two spectacle makers in Holland, Jans and Zacharias Jansen. The lenses were crude and the magnification achieved was accordingly unimpressive. But in 1683 Anton van Leeuwenhoek, also of Holland, using better lenses and a specimen carrier, brought the magnification up to 275X. With his microscope, Leeuwenhoek discovered living microorganisms such as infusoria in water and spirilla in particles between the teeth. In 1850, Giovanni Battista Amici of Florence became famous for the high quality of his microscopes, having improved the lens system to correct color distortions by means of achromatic lenses. Around 1868, Carl Zeiss, the owner of an optical workshop, improved the quality of lens grinding through precise mathematical computations of lens properties. Ernst Abbe became a member of his company and, after years of experimentation, perfected objectives of outstanding quality—apochromatic lenses—which, with compensating eyepieces, introduced a new era in microscope objective design. Under the guidance of Frits Zernike (a Nobel Prize winner), Zeiss developed the first phase contrast microscope, which permitted the viewing of transparent objects, such as living cells—a monumental aid to biological research. Objectives were further improved by correcting them for a flat field, permitting sharpness of the field from center through periphery. These refinements led to the development, in 1957, of the planapochromats, the finest lenses available for photomicrography. With the development of sophisticated lighting systems, vertical illuminators, condensers and polarizing equipment, the microscope has become an indispensable tool in medicine, criminal investigation, biology, chemical analysis, geology, industry and scientific research. The variety of design and form that it reveals makes it an exciting artistic tool as well.

Photography through a microscope is relatively easy. After removing the lens of a 35mm camera, the camera, set at infinity, is either directly attached to the eyepiece of the microscope through a light-tight tube or sleeve adapter, or, in the case of heavier cameras with a larger negative format, fixed to a separate stand and attached to the eyepiece by a light-tight adapter. Thus the lens of the microscope becomes in effect a functioning lens of extremely short focus. Objective lenses of the microscope vary in the degree of magnification they provide—2.5X, 4X, 10X, 16X, 25X, 63X, 100X, the markings appearing on the barrel of the lens. The eyepiece lenses magnify the enlarged image of the eyepiece by the magnification factors marked on them, 8X or 10X, for example. Thus a 10X objective lens used with a 8X eyepiece lens provides a magnification of 80X. If the developed negative is further enlarged in making the final print, the lens magnification is multiplied by the degree of enlargement. A 5X enlargement of a negative made with an 80X magnification will provide a final image of 400X. Objectives of 4X, 10X and 16X magnifications are most often employed for pictorial purposes since they yield great depth of focus and extremely sharp pictures.

Focusing is done through the rack and pinion arrangement on the microscope which usually provides for both rough and fine adjustments. It is critically important to focus the image sharply at the plane of the negative. This is a comparatively simple procedure with a single lens reflex camera. Once the image is brought into focus, the prism of the camera is removed and the ground glass viewed through a simple magnification lens (I use a

10X lens). The microscope is then readjusted for the sharpest focus.

The illumination unit in photomicrography must provide intense concentrated lighting. This is most conveniently supplied by a 500-watt-filament, gas-filled tungsten lamp which is either directed at a tilted mirror underneath the specimen stage to reflect the light onto the object to be viewed, or, in the more expensive microscopes, is built into the base in a lamp housing. A field condenser and a diaphragm are part of the unit and the light is beamed into the microscope substage condenser which distributes it evenly throughout the specimen.

Since the light floods the specimen from below, the specimen must obviously be thin enough to permit the penetration of light beams into the lens system. This is called bright-light illumination, and it is employed for the viewing of slides. Slide preparation involves immersing soft objects such as plants and animal tissues into chemicals to harden them, dehydrating them in absolute alcohol, impregnating them in molten paraffin and running the hardened paraffin block through a slicing tool (microtome) which shaves off a thin slice that is then mounted on a glass slip.

In the early stages of preparation, specimens are often treated with various stains. Since different parts of cells or substances pick up certain stains while rejecting others, these stains can intensify dramatically the structural composition of many objects. Some objects require no stains, for contrasting parts can often be shown clearly by such other techniques as oblique lighting—the slight displacement of the light beam to illuminate the object from the side. Colorless objects are sometimes effectively seen through dark-field illumination promoted by a special dark-field sub-stage condenser which sets off the tissues in white against a black background. Translucent objects may also be viewed successfully through phase-contrast or interference-contrast lighting systems which require special condensers and objectives. Some objects, such as rocks, require no advanced chemical treatment or staining and are sliced by a special microtome into paper-thin sections for viewing. Chemicals to be viewed are merely dissolved in water and then dropped on a glass slide; the translucent crystals are then viewed directly or through polarizing filters, one placed above the specimen and one below it. Rotation of one of the filters produces dramatic color effects which change as the positions of the filter vary. Rocks, chemical crystals, hairs, and certain tissues (even those that have been stained) may produce startling effects when viewed through polarization devices. It stands to reason that opaque objects such as metals require illumination from above rather than through the bottom of the stage. Top lighting is provided by a special vertical illuminator and distinctive (epiplan) lenses.

Through the use of the electron microscope direct magnifications of up to 30,000 diameters may be obtained which may be further optically enlarged to yield magnifications of 100,000 diameters or more. Thus the electron microscope permits the examination of objects as small as viruses and the ingredients of chromosomes.

The pictures in this volume are representative of the wide range of natural forms available for artistic use and enjoyment. They were all made with a

Zeiss Standard WL Microscope. The objective lenses used for bright-field illumination were of the plano-apoachromatic variety. For the photography of metals, a Zeiss vertical illuminator was employed with epiplan objective lenses. A selective search for the most flexible camera narrowed the choice to the Rollerflex SL66, a single-lens reflex camera with a picture size of 2¼″ x 2¼″. The exposures were calculated with a Gossen Lunasix light meter. Black-and-white films were limited to Panatomic X (which has recently been discontinued) and Plux X. Agfachrome Professional 50 and Agfachrome CT 18 were used for the color transparencies.

LEWIS R. WOLBERG, M.D.

New York City, 1974

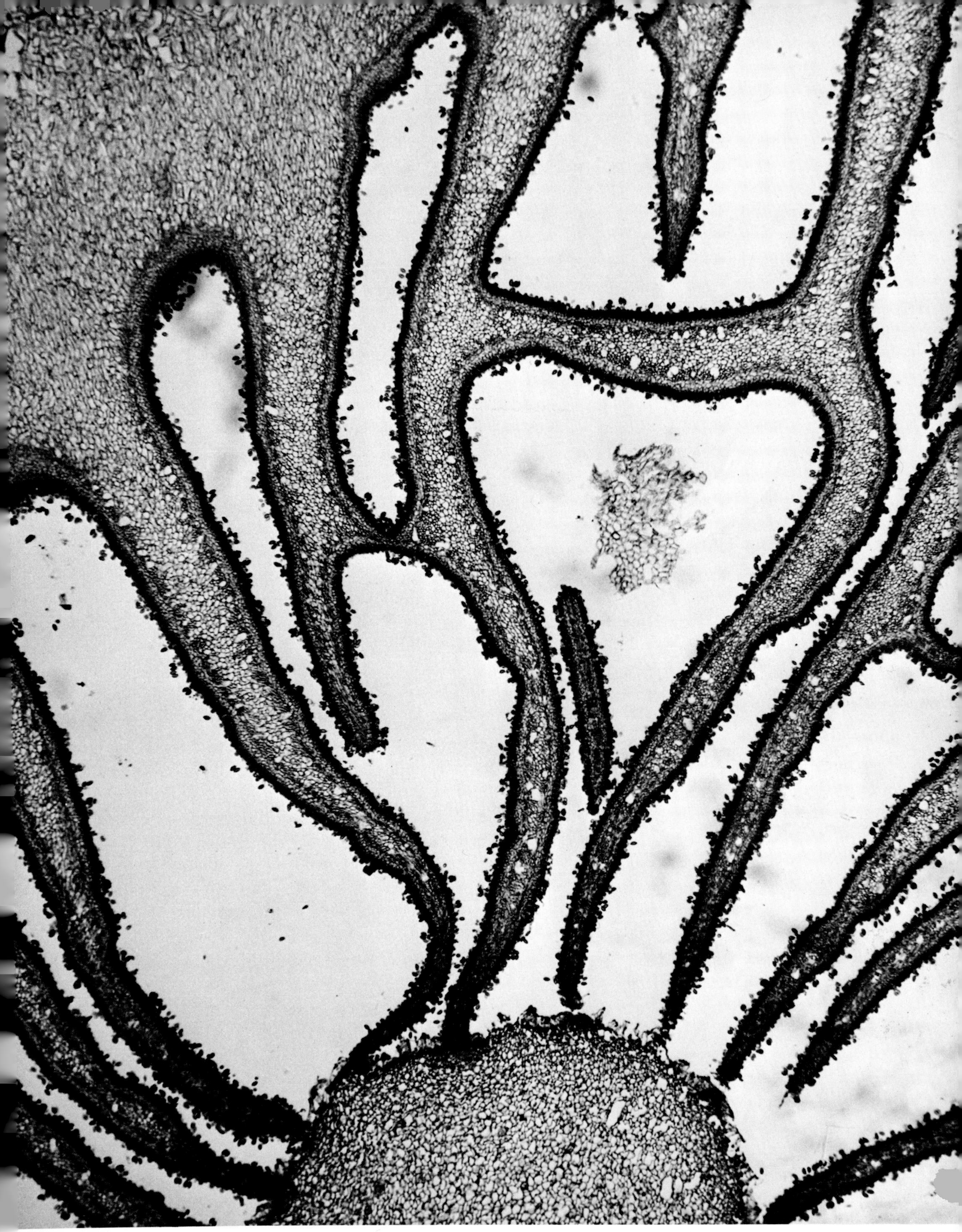

1 Fungus *Coprinus*. Magnification 175X.

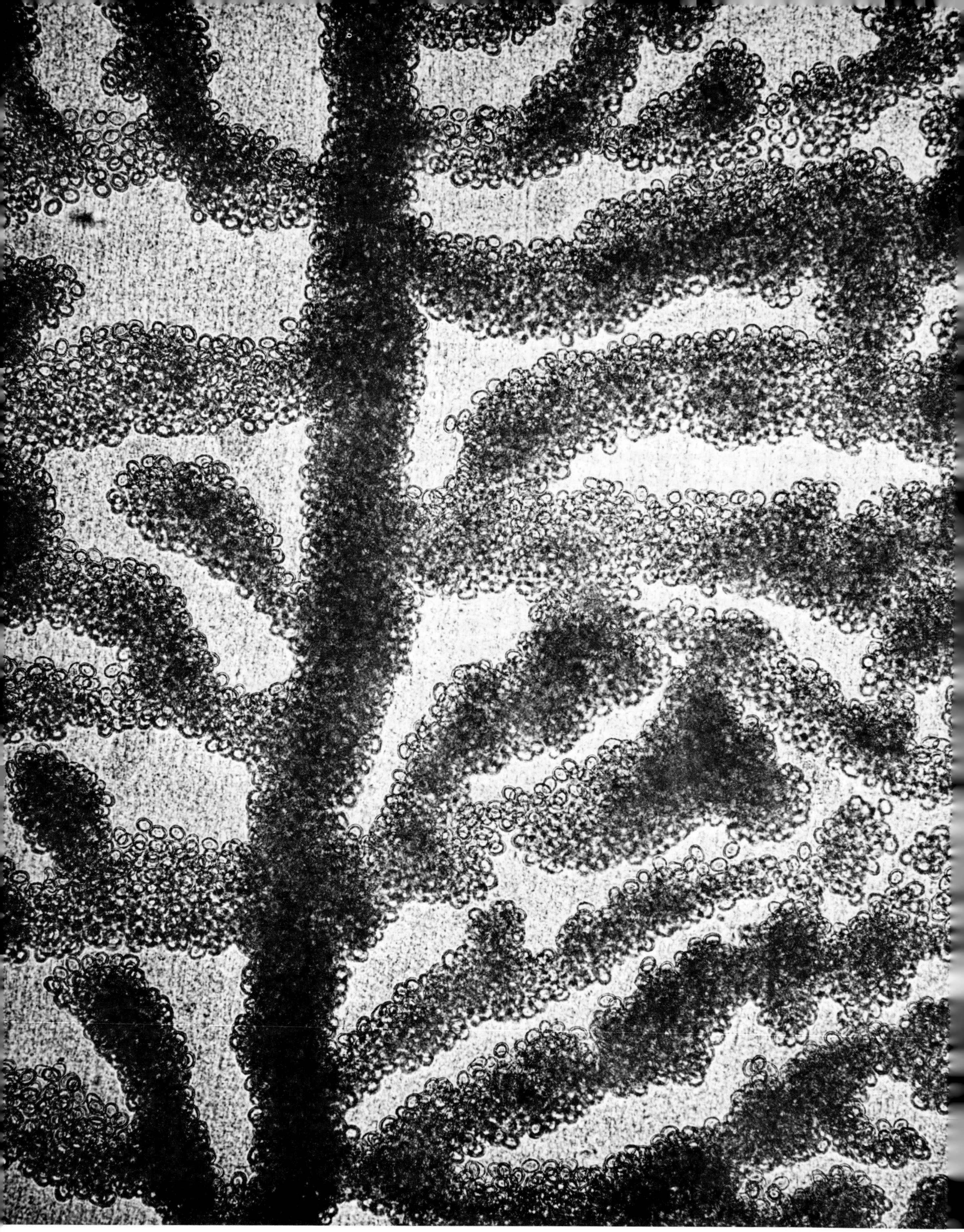

2 Tapeworm *(Taenia pisiformis)*. Magnification 175X.

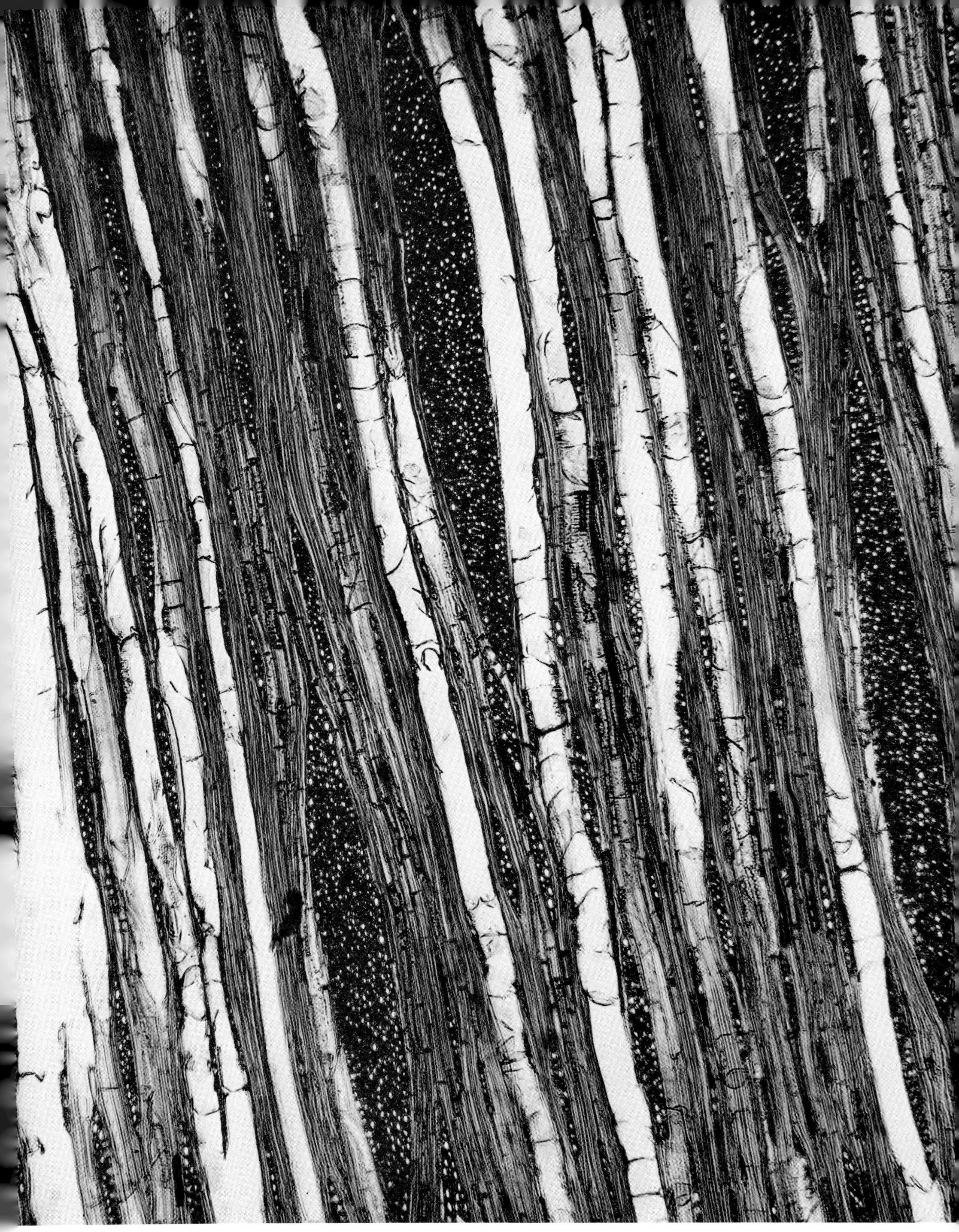

3 Beechwood. Magnification 210X.

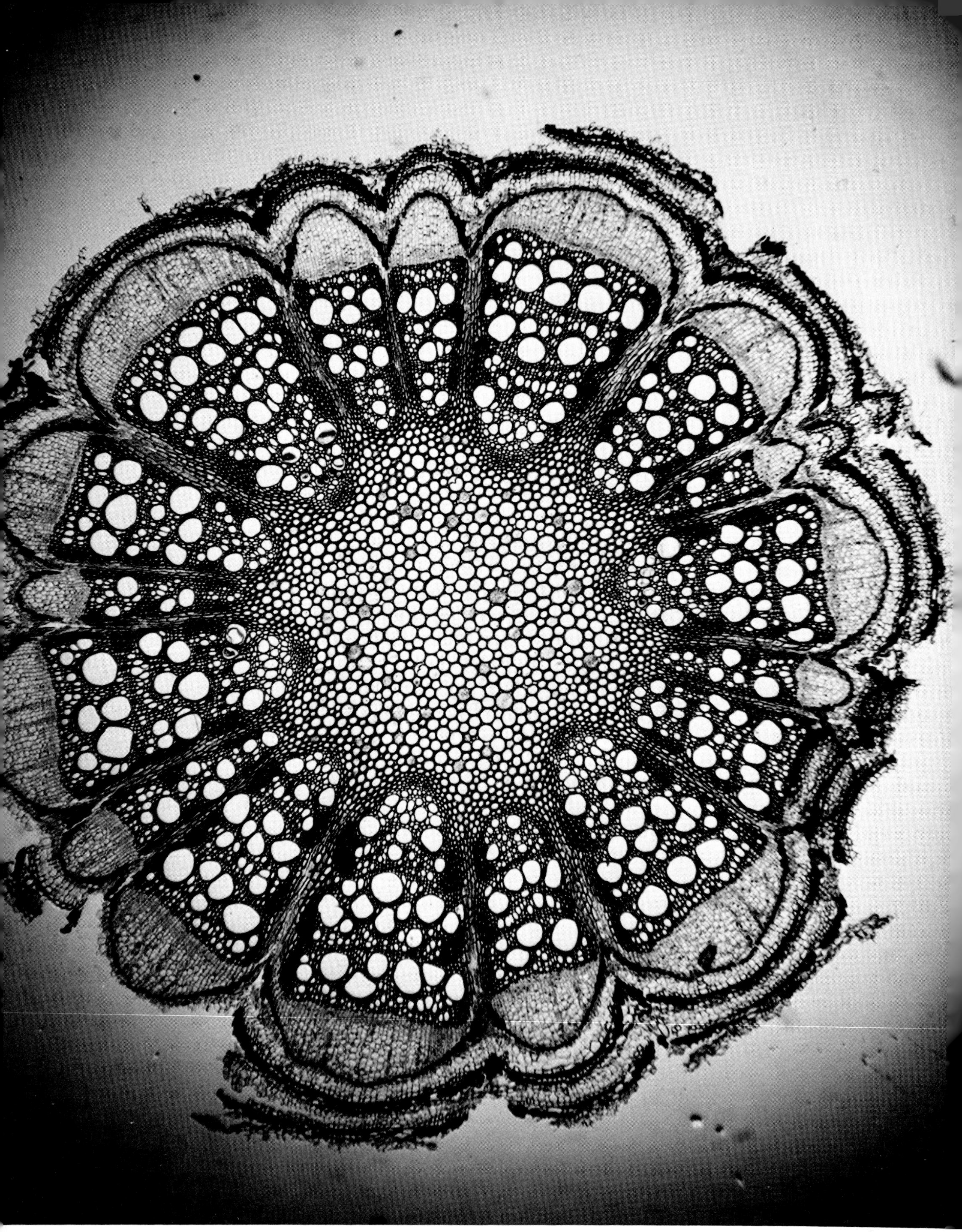

4 Clematis stem. Magnification 110X.

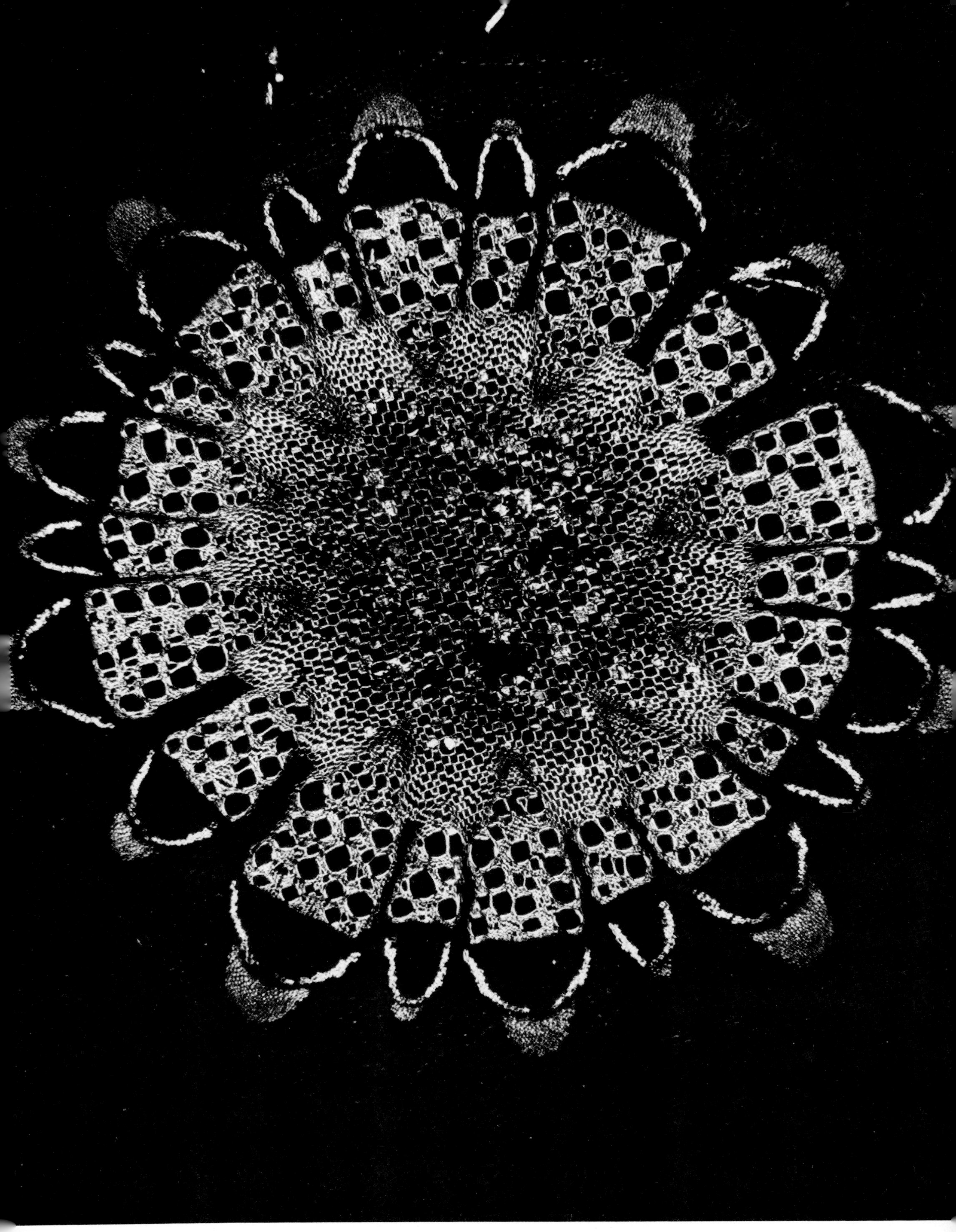

5 Clematis stem. Magnification 85X.

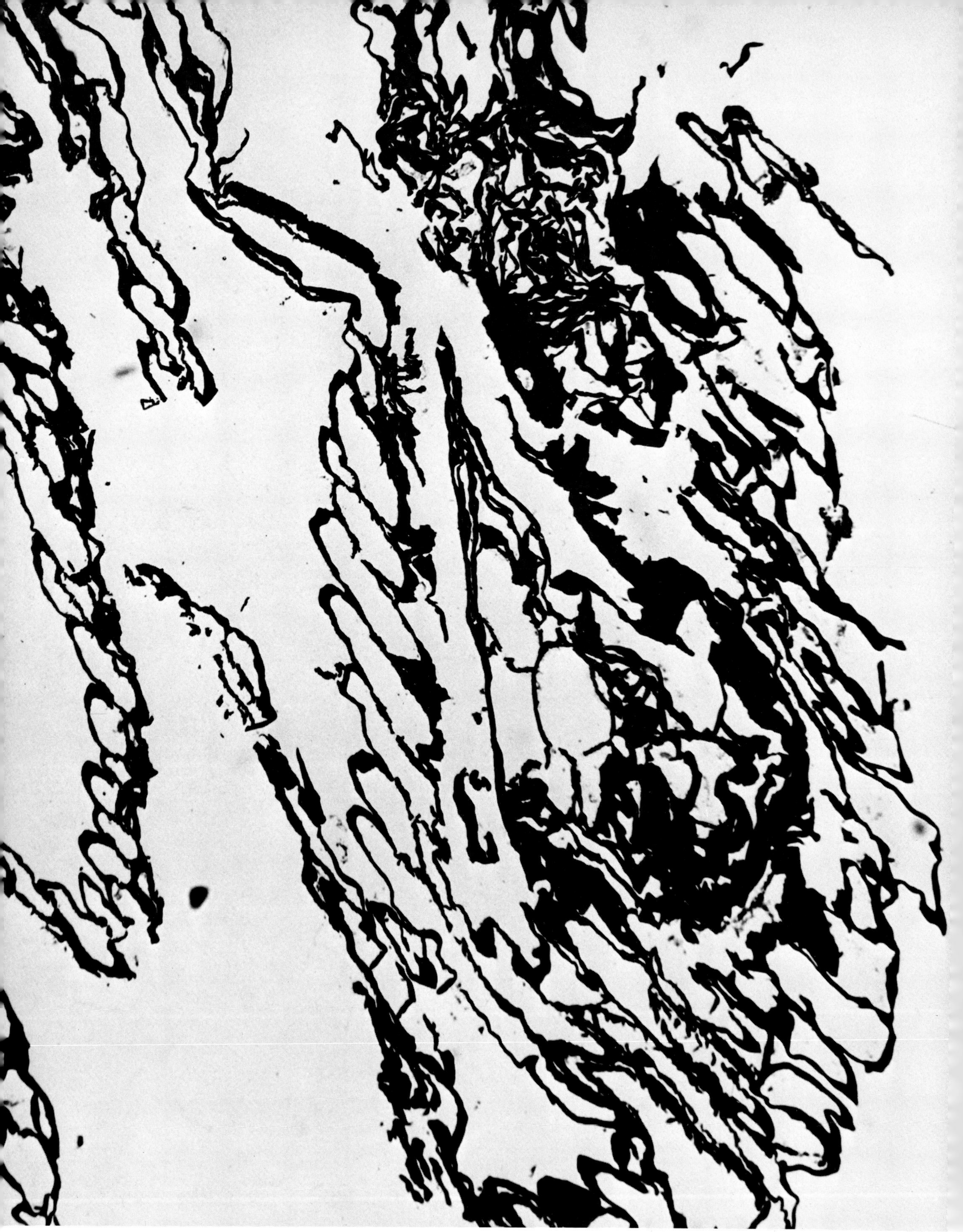

6 Onion skin. Magnification 560X.

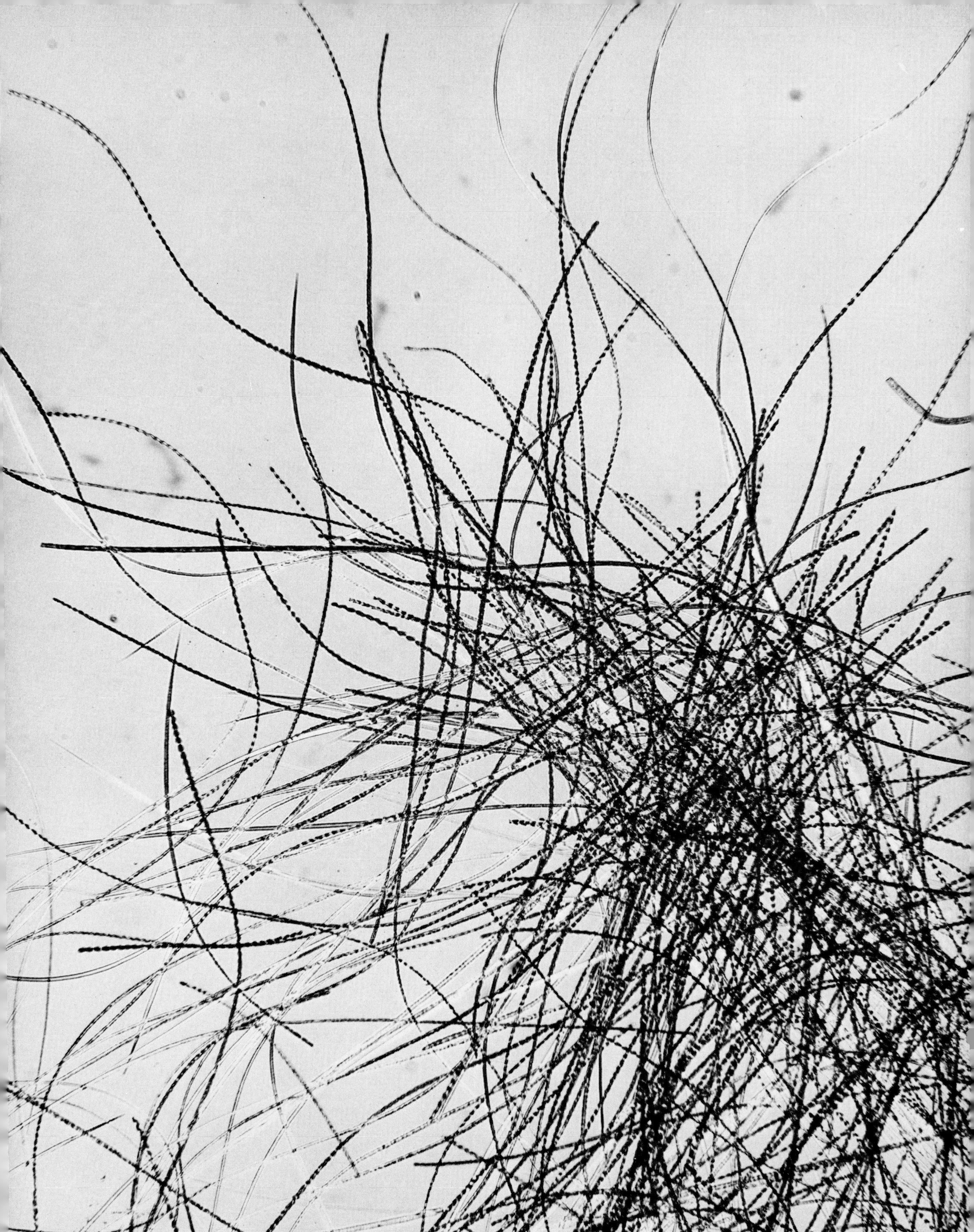

7 Bat hairs. Magnification 155X.

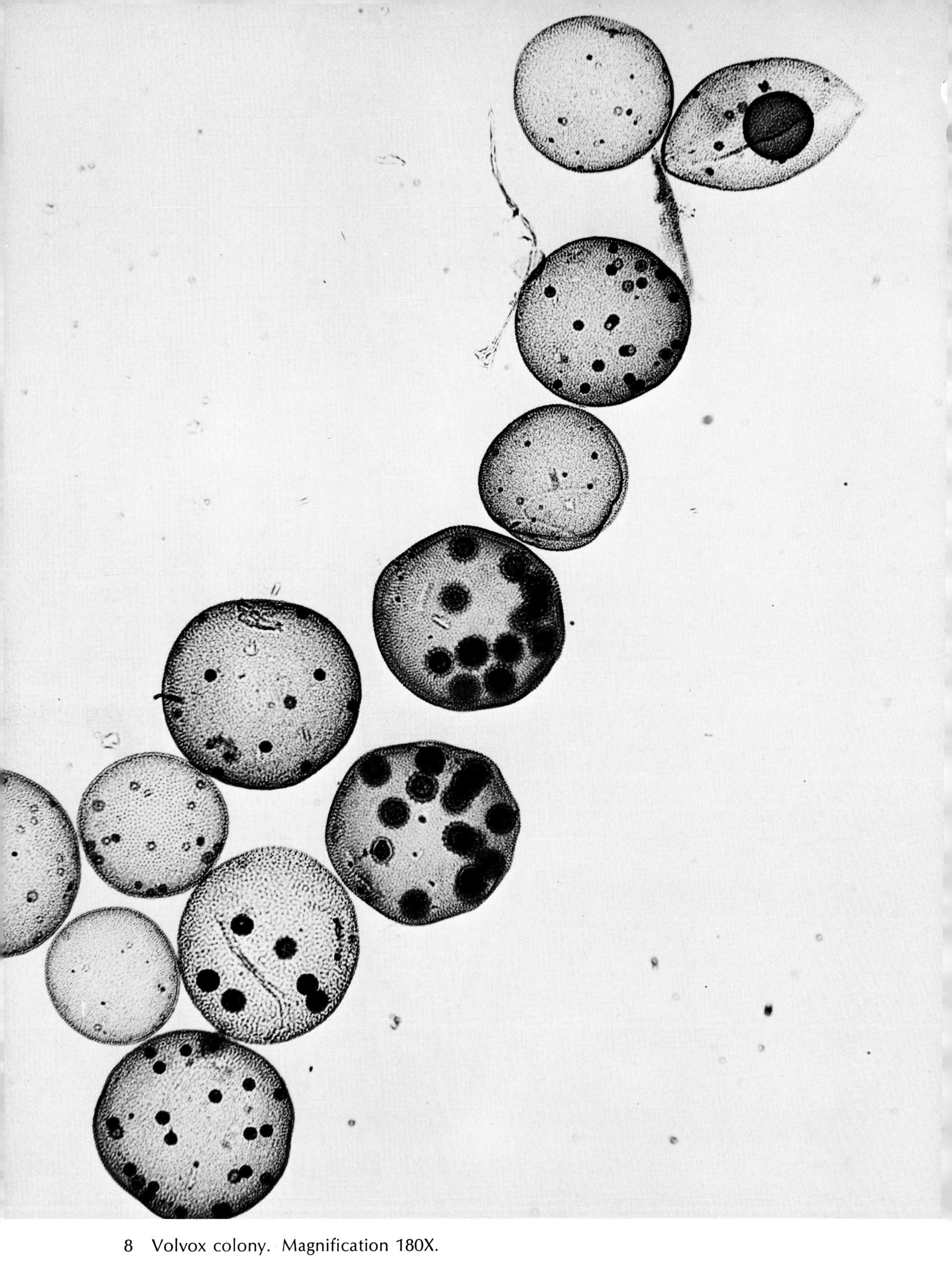

8 Volvox colony. Magnification 180X.

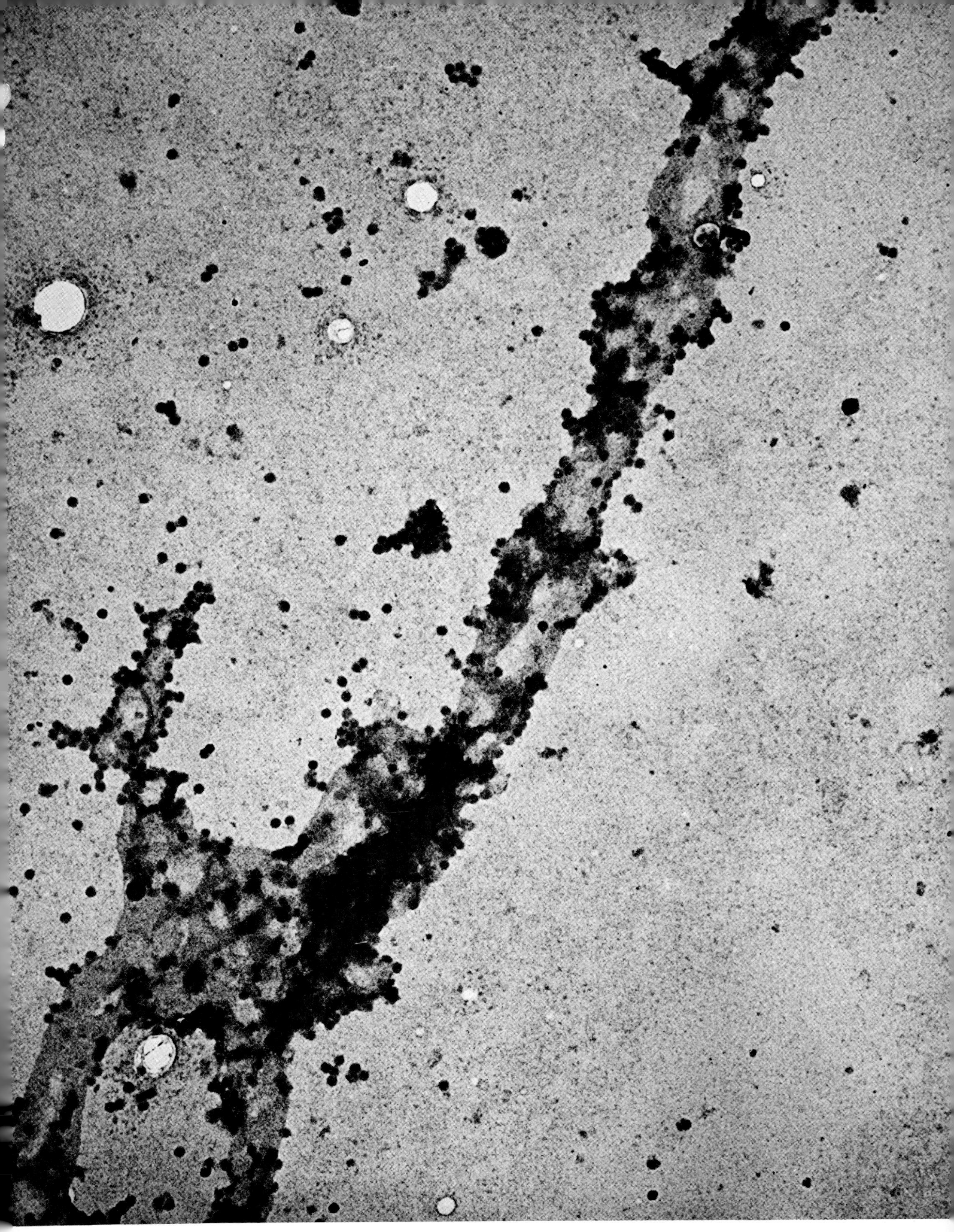

9 Mayaro virus. Magnification 140,000X.

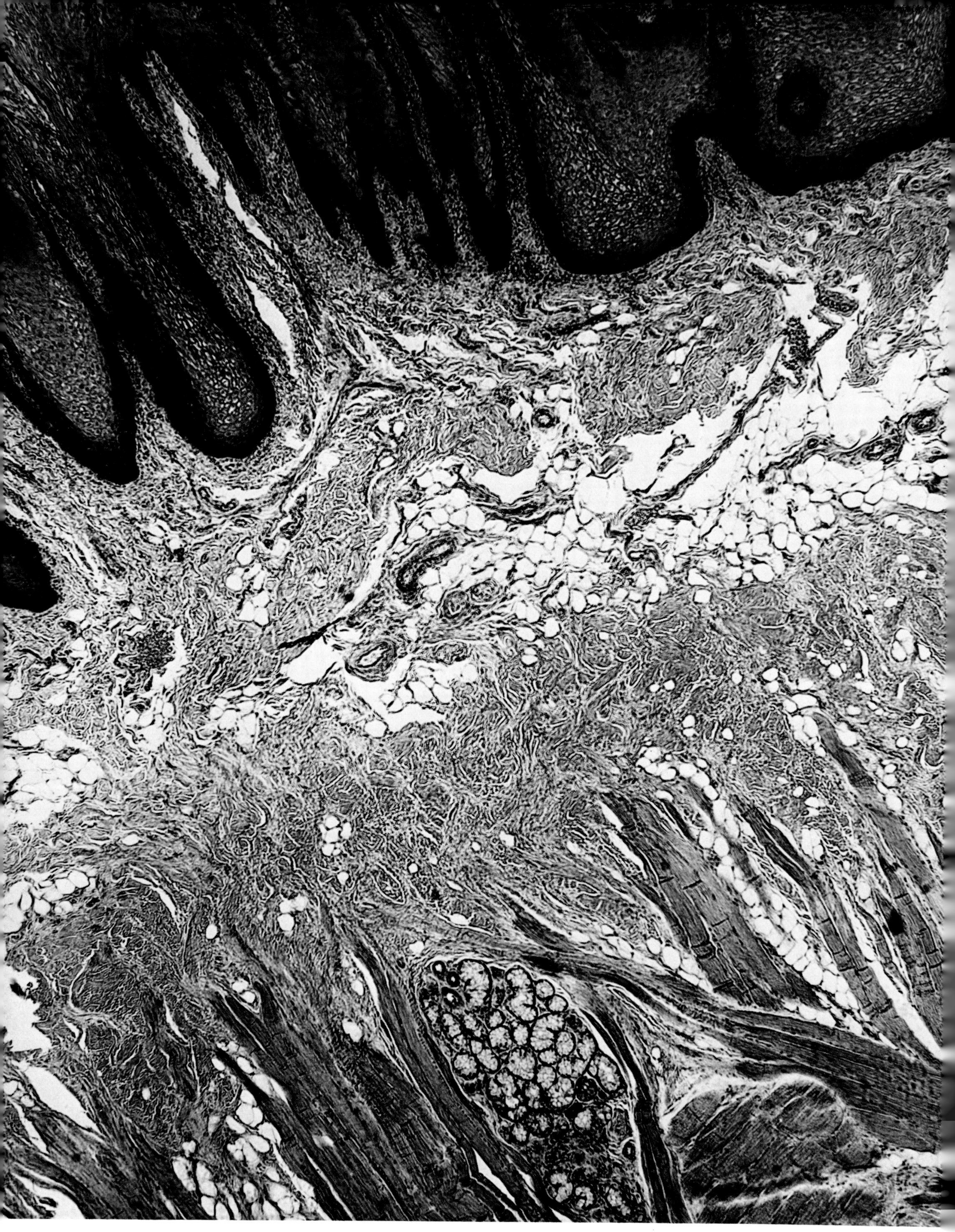

10 Human tongue. Magnification 175X.

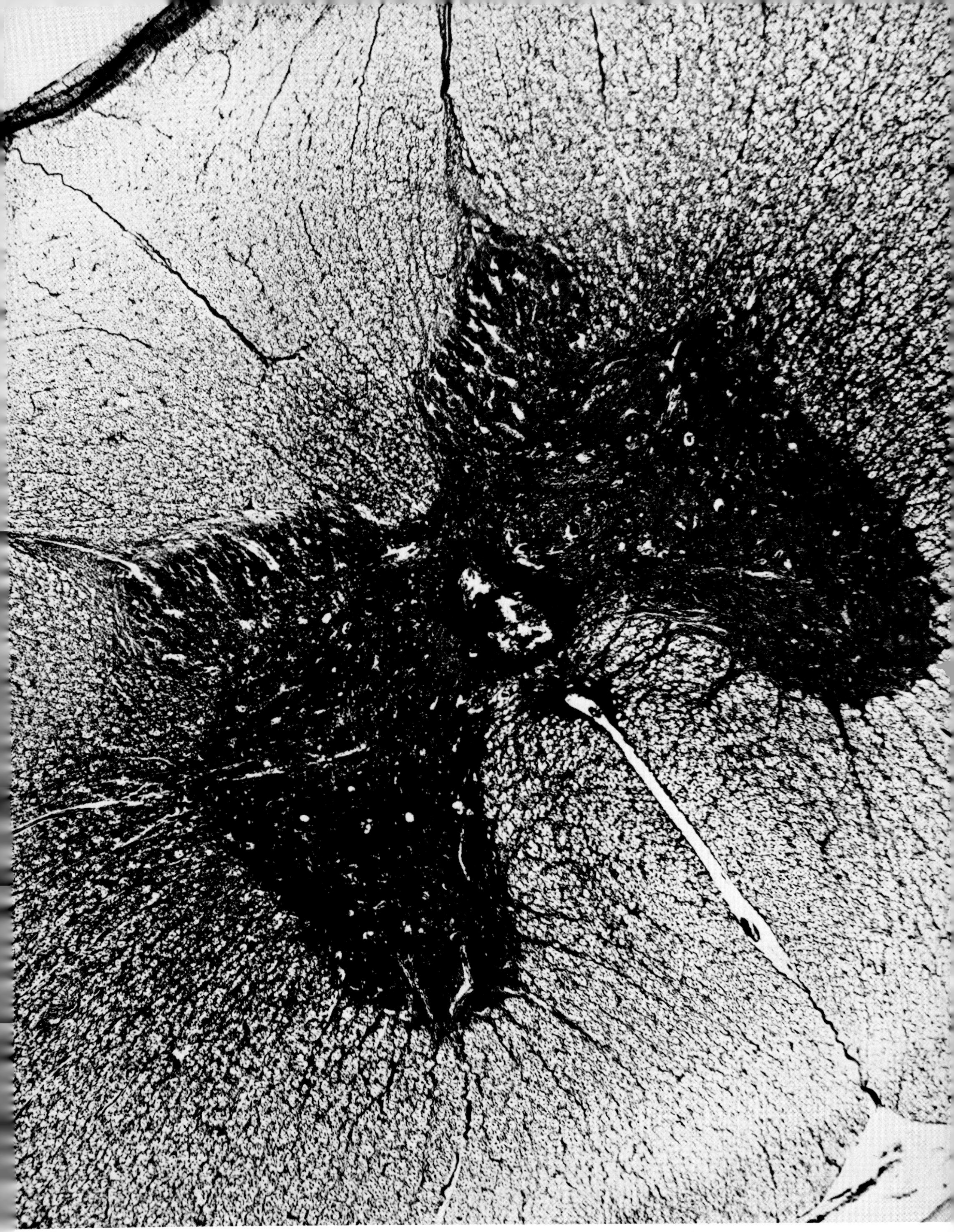

11 Human spinal cord. Magnification 175X.

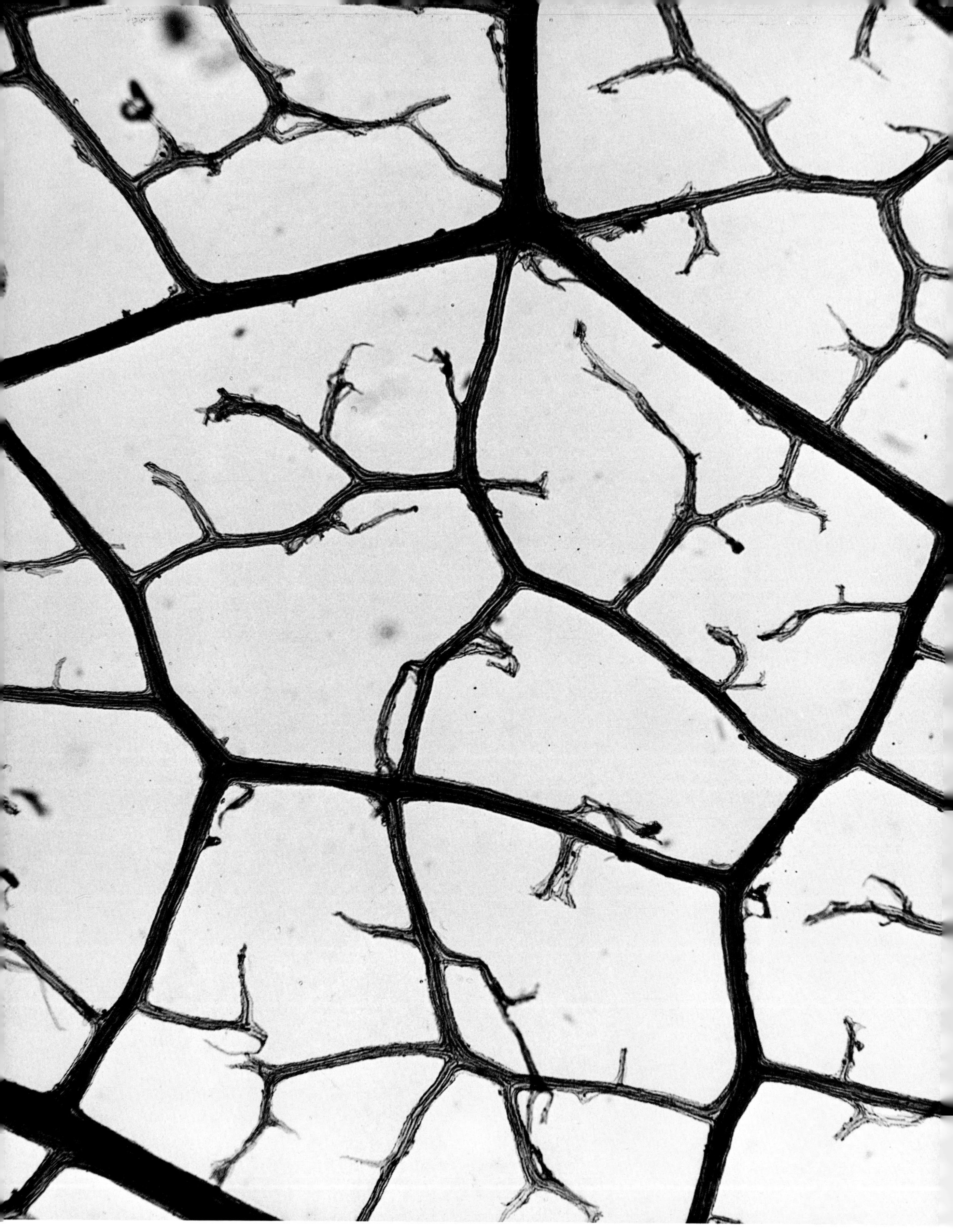

12 Leaf skeleton. Magnification 175X.

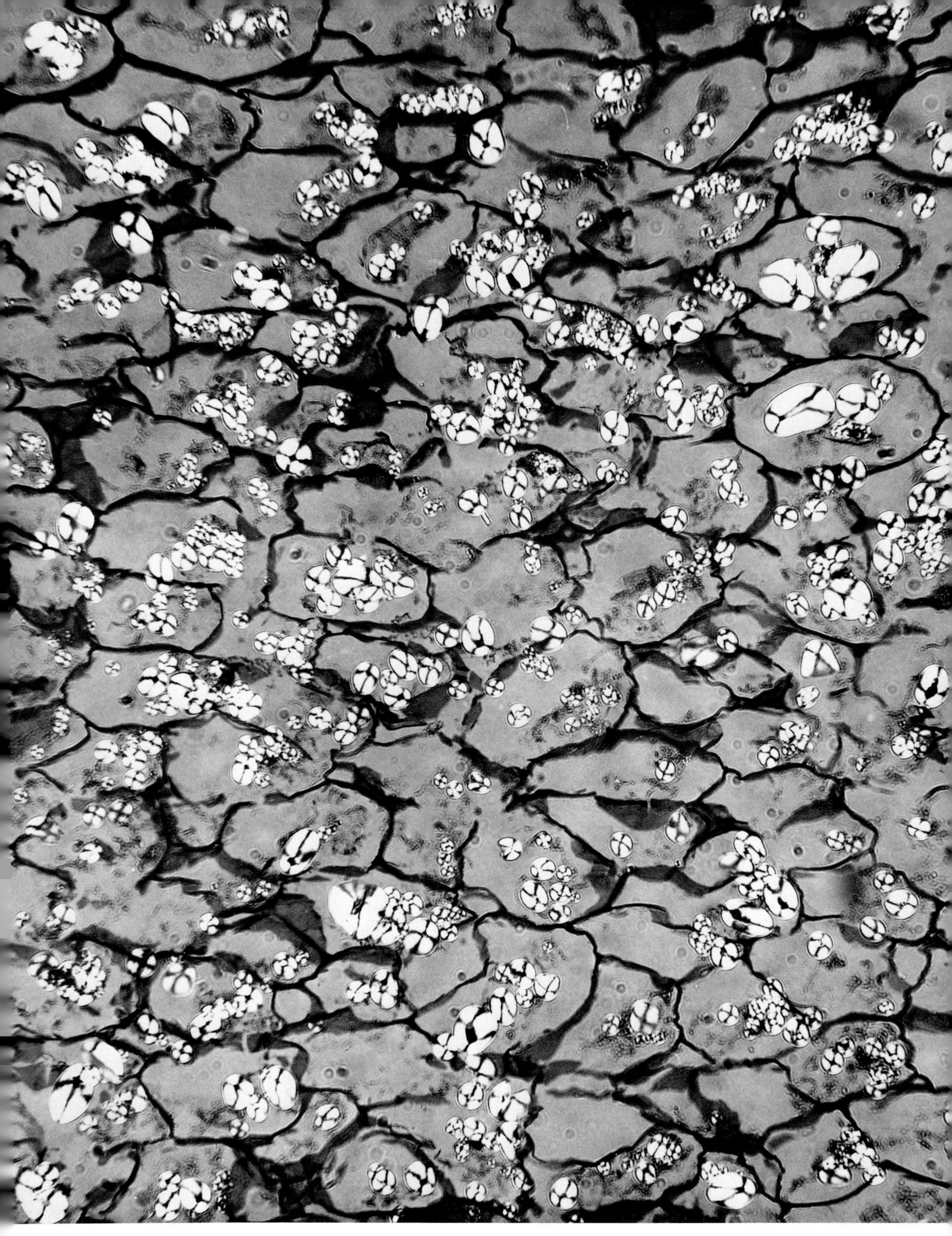

13 Potato scab. Magnification 420X.

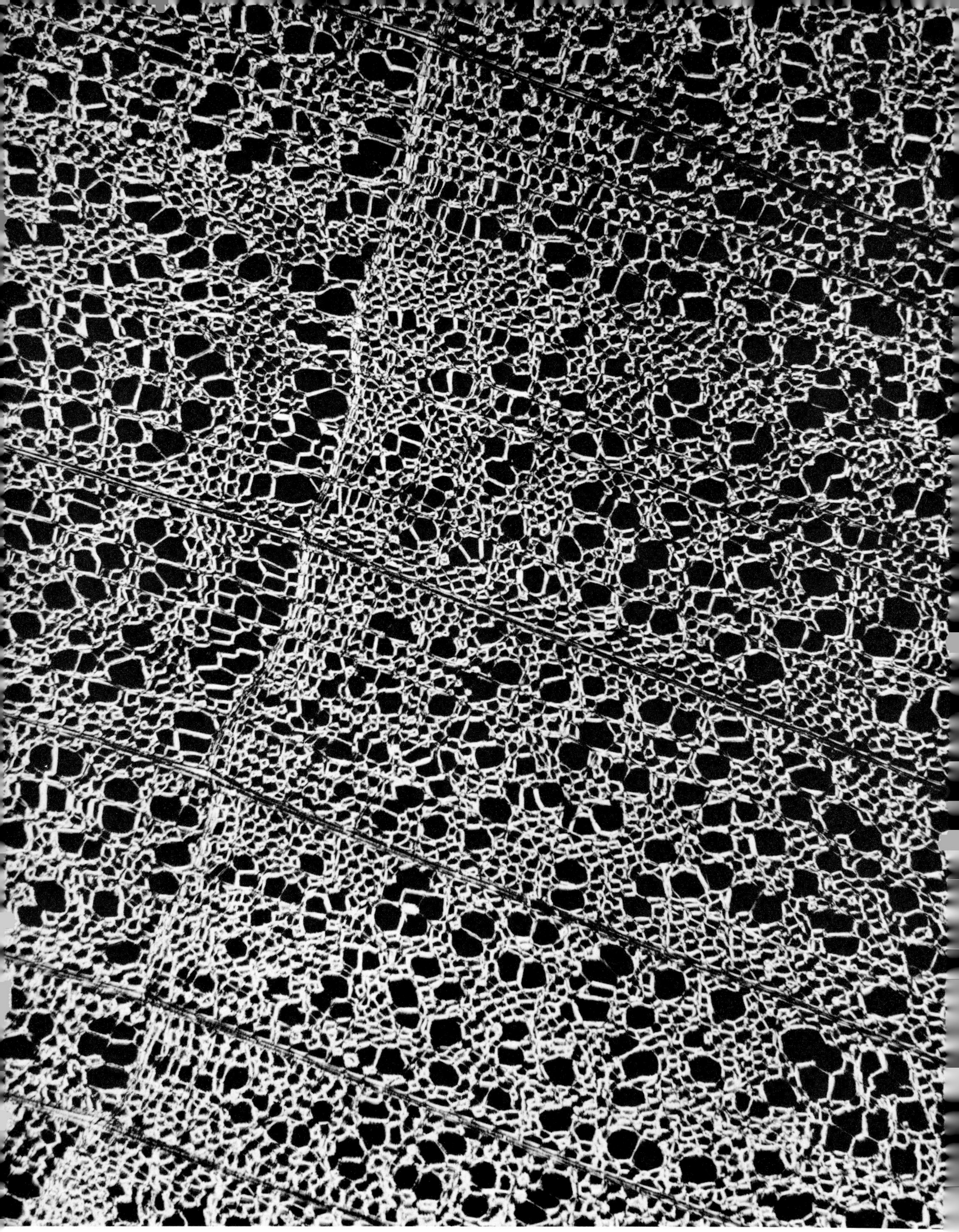

14 Basswood. Magnification 170X.

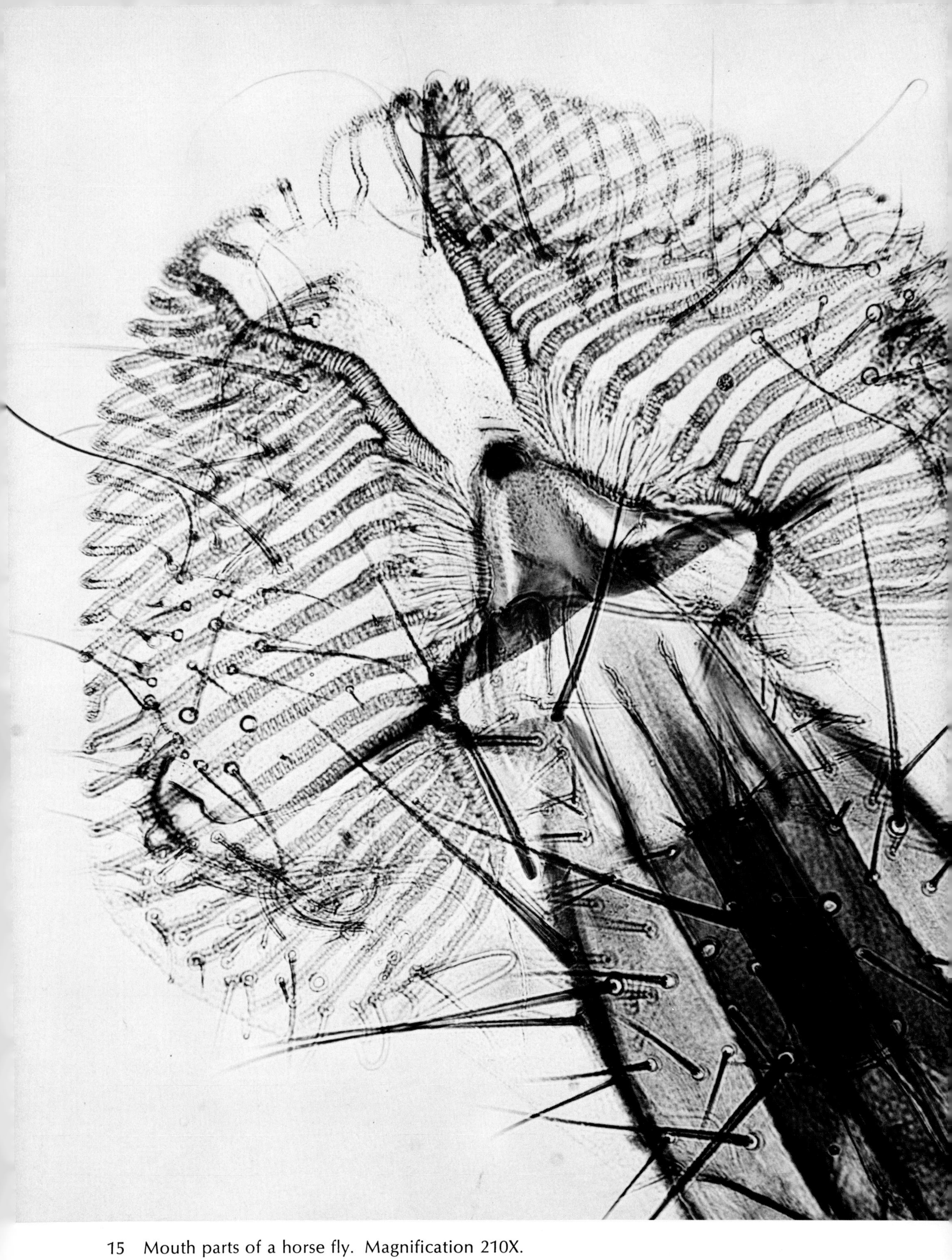

15 Mouth parts of a horse fly. Magnification 210X.

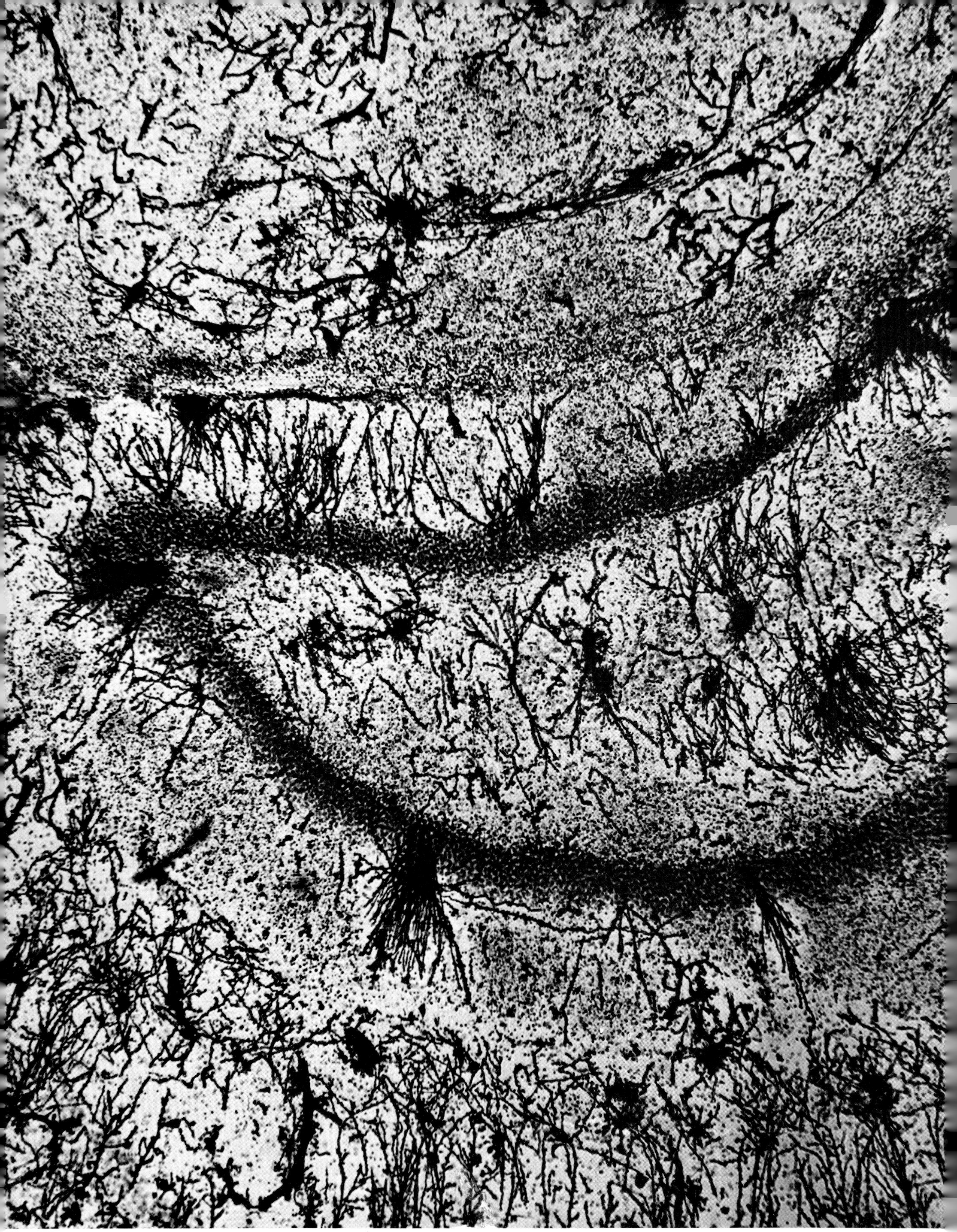

16 Cells from the frontal lobe of a rat's brain. Magnification 210X.

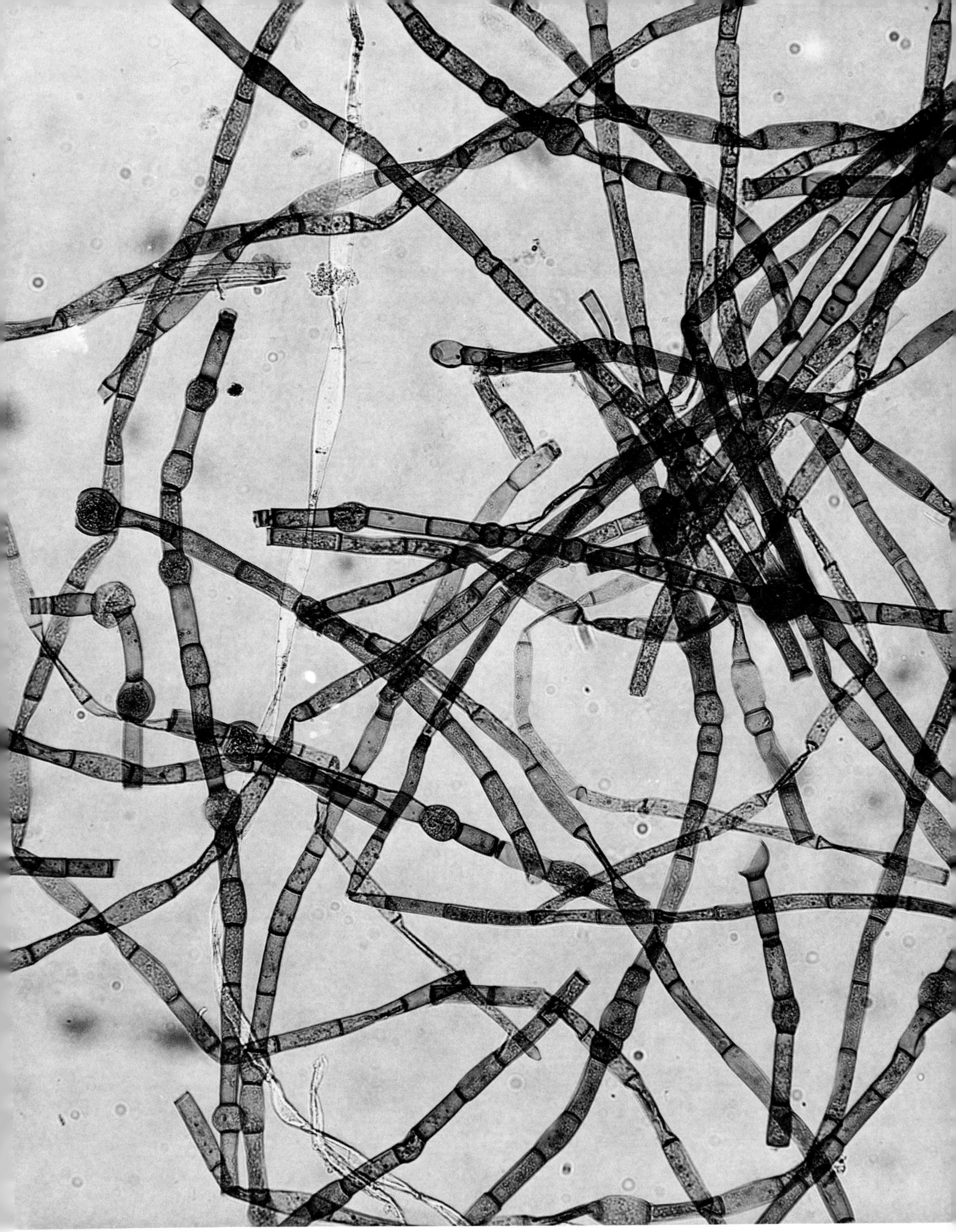

17 Algae *Oedogonium*. Magnification 450X.

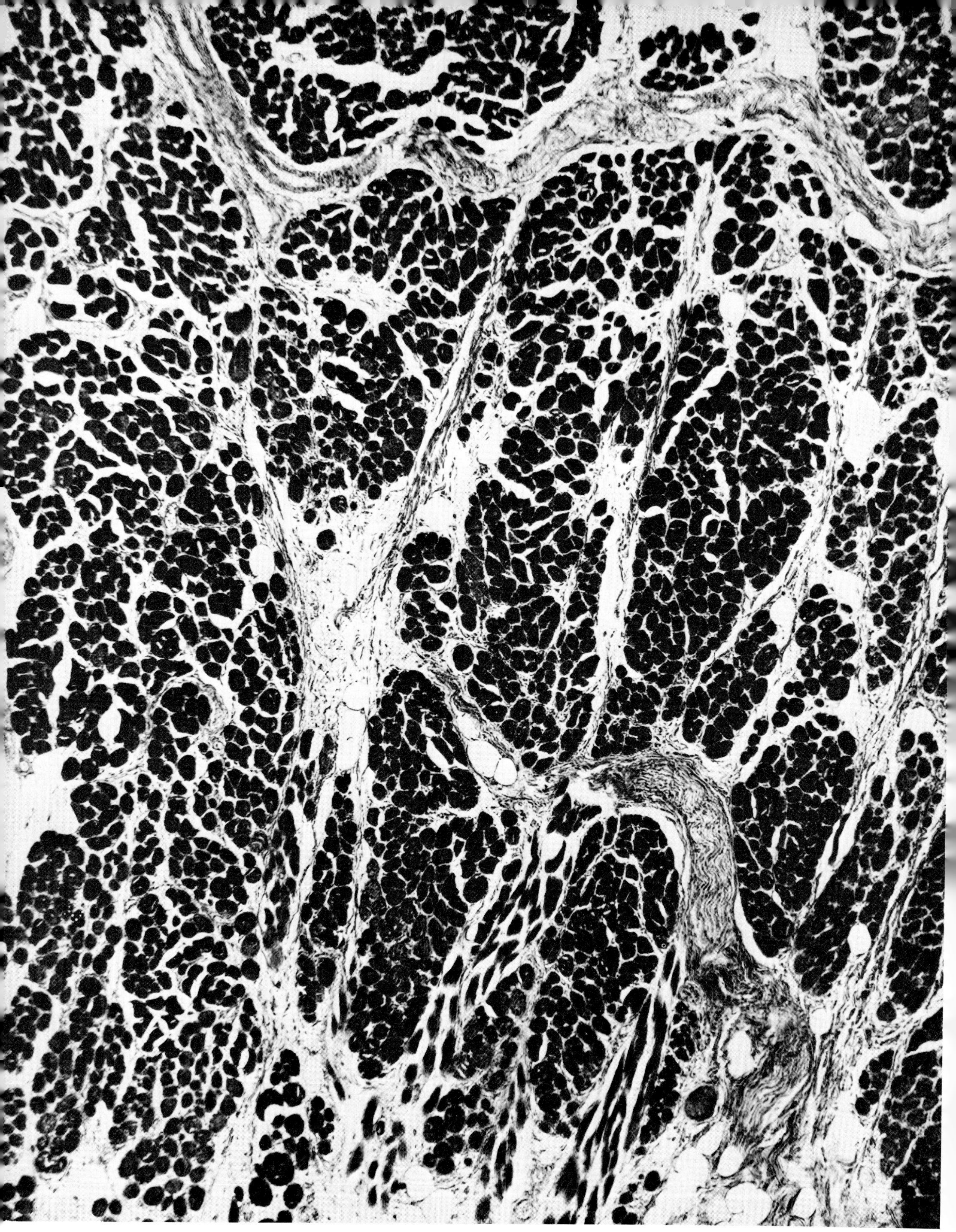

18 Human lip. Magnification 210X.

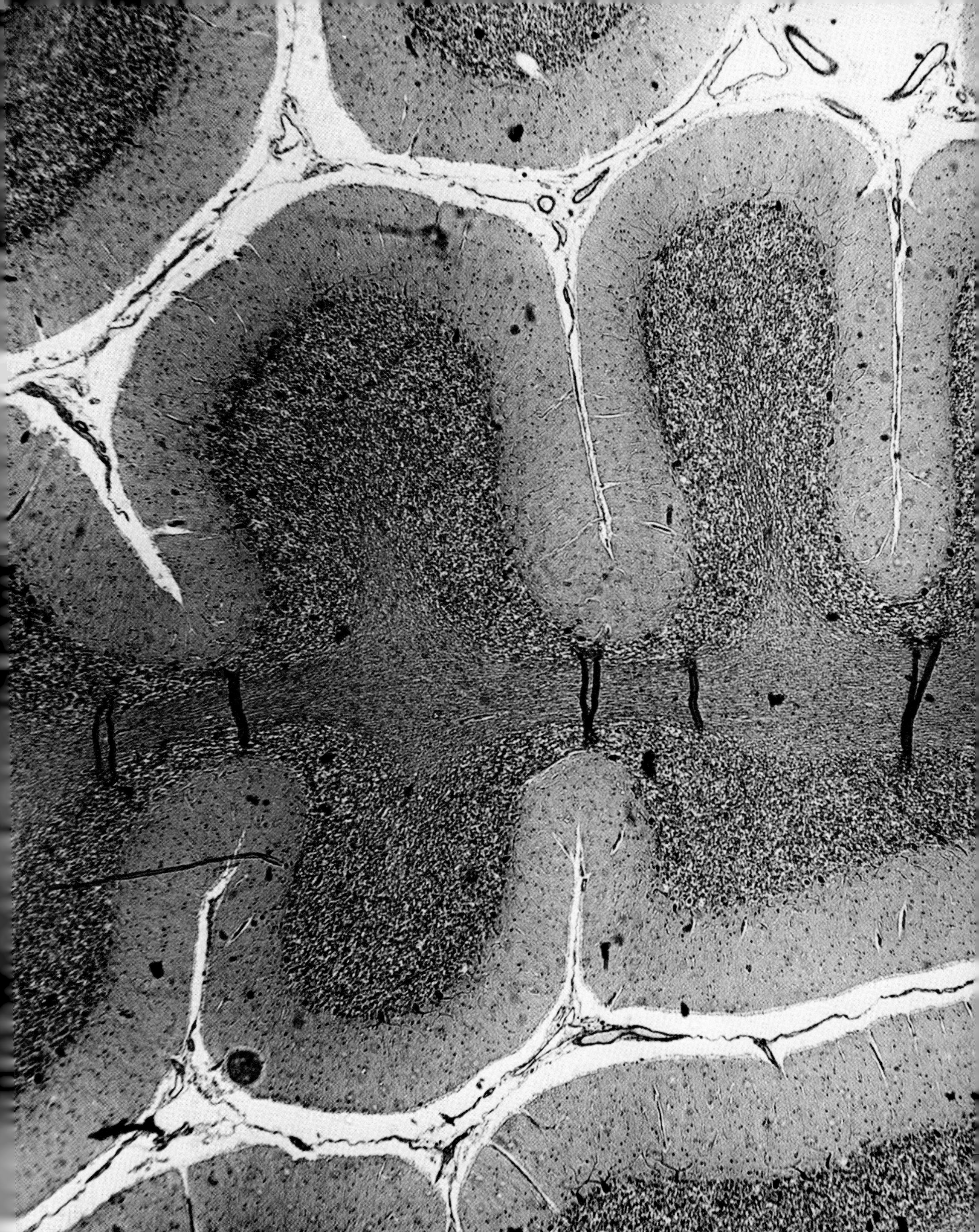

19 Human cerebellum. Magnification 125X.

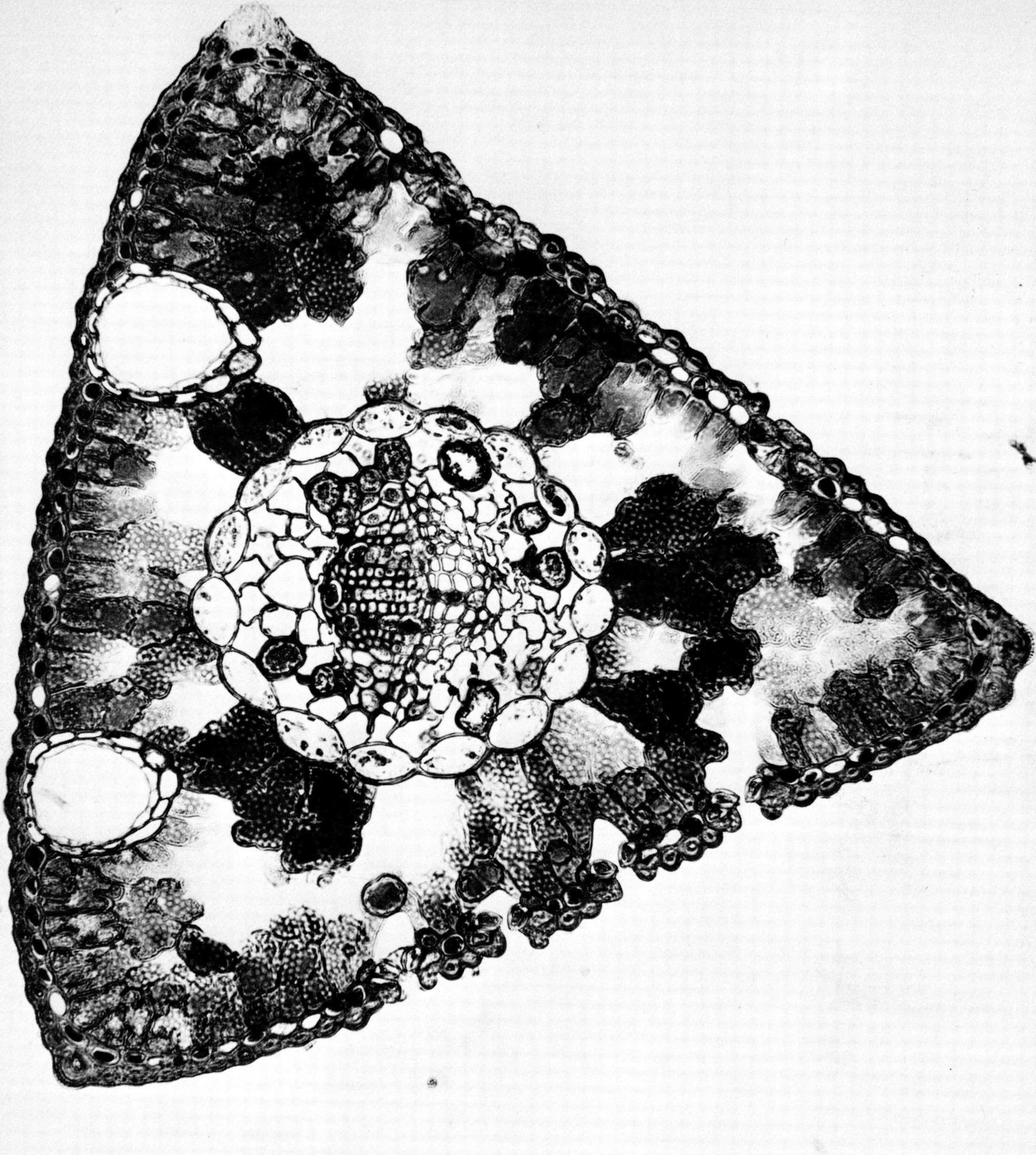

20 Leaf of Pinus Strobus. Magnification 450X.

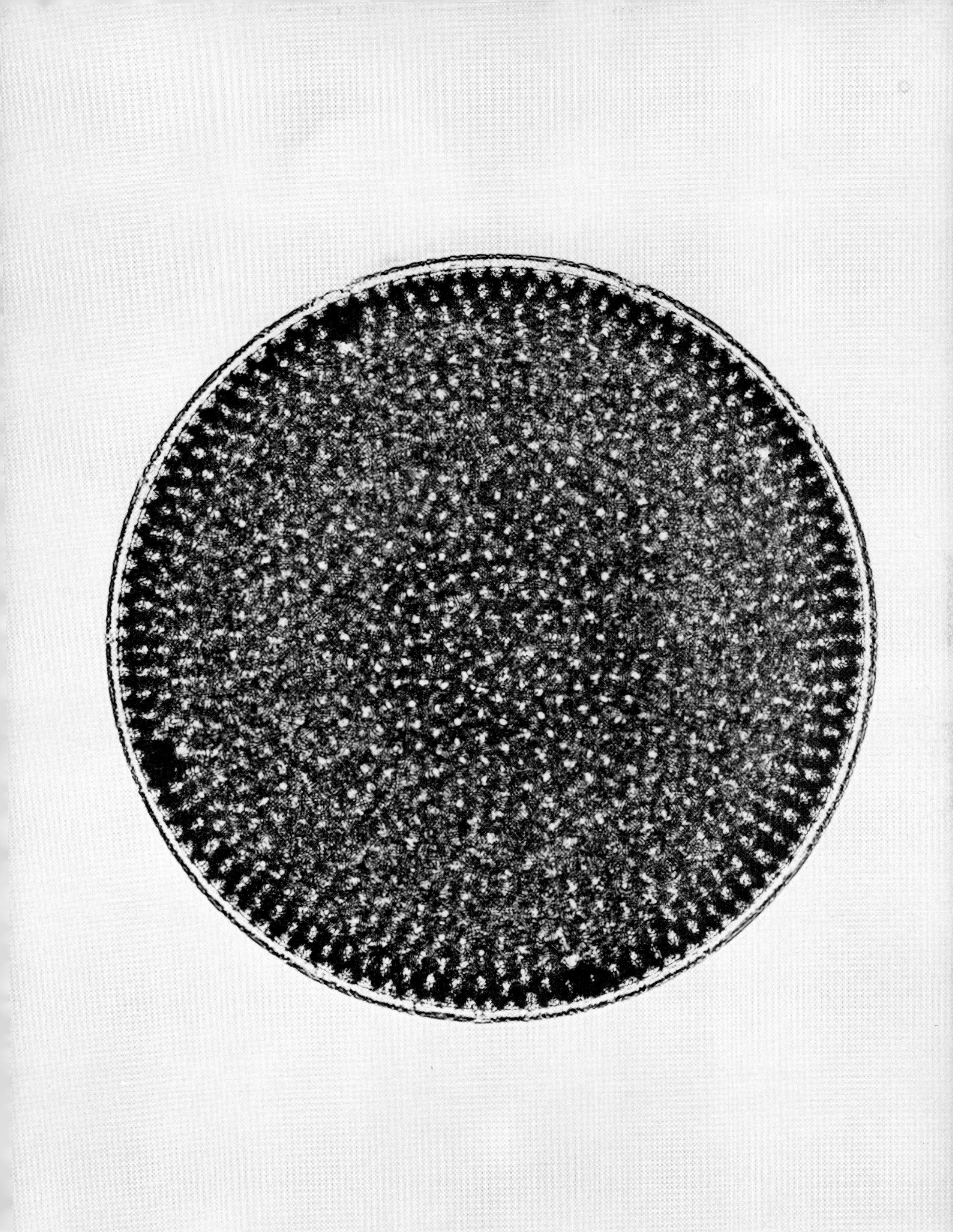

21 Diatom. Magnification 1050X.

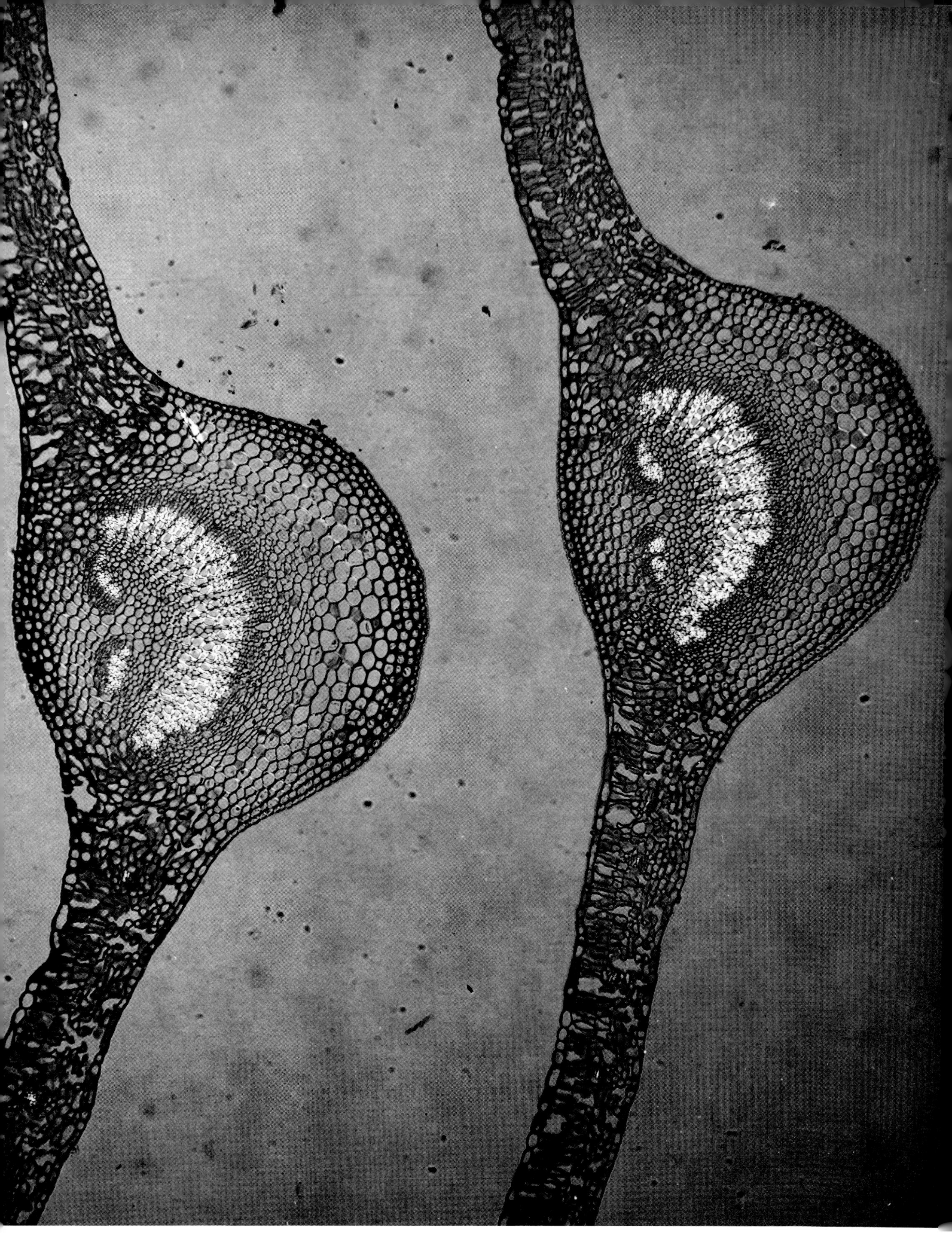

22 Lilac leaves. Magnification 175X.

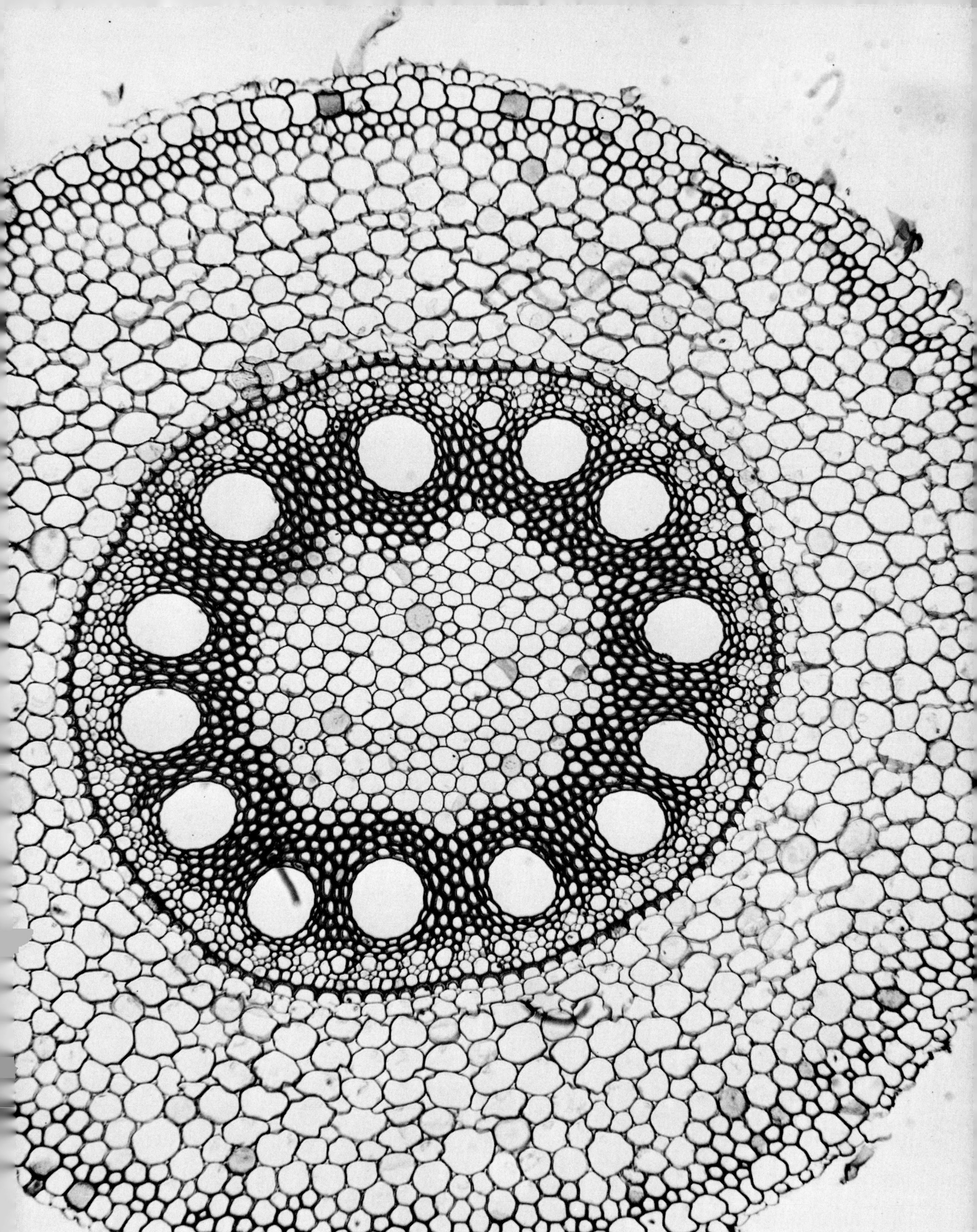

23 Corn root. Magnification 350X.

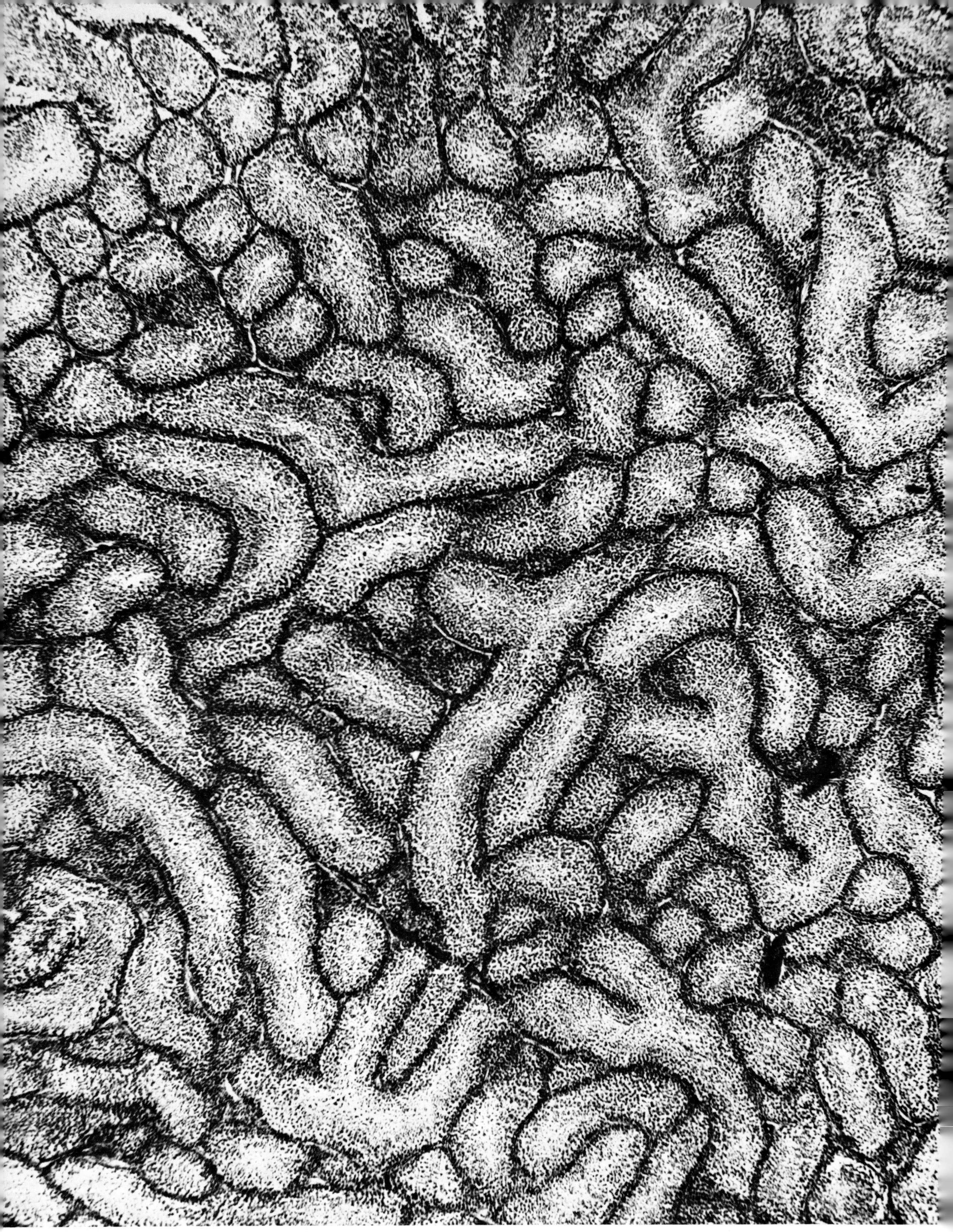

24 Testes of dove. Magnification 455X.

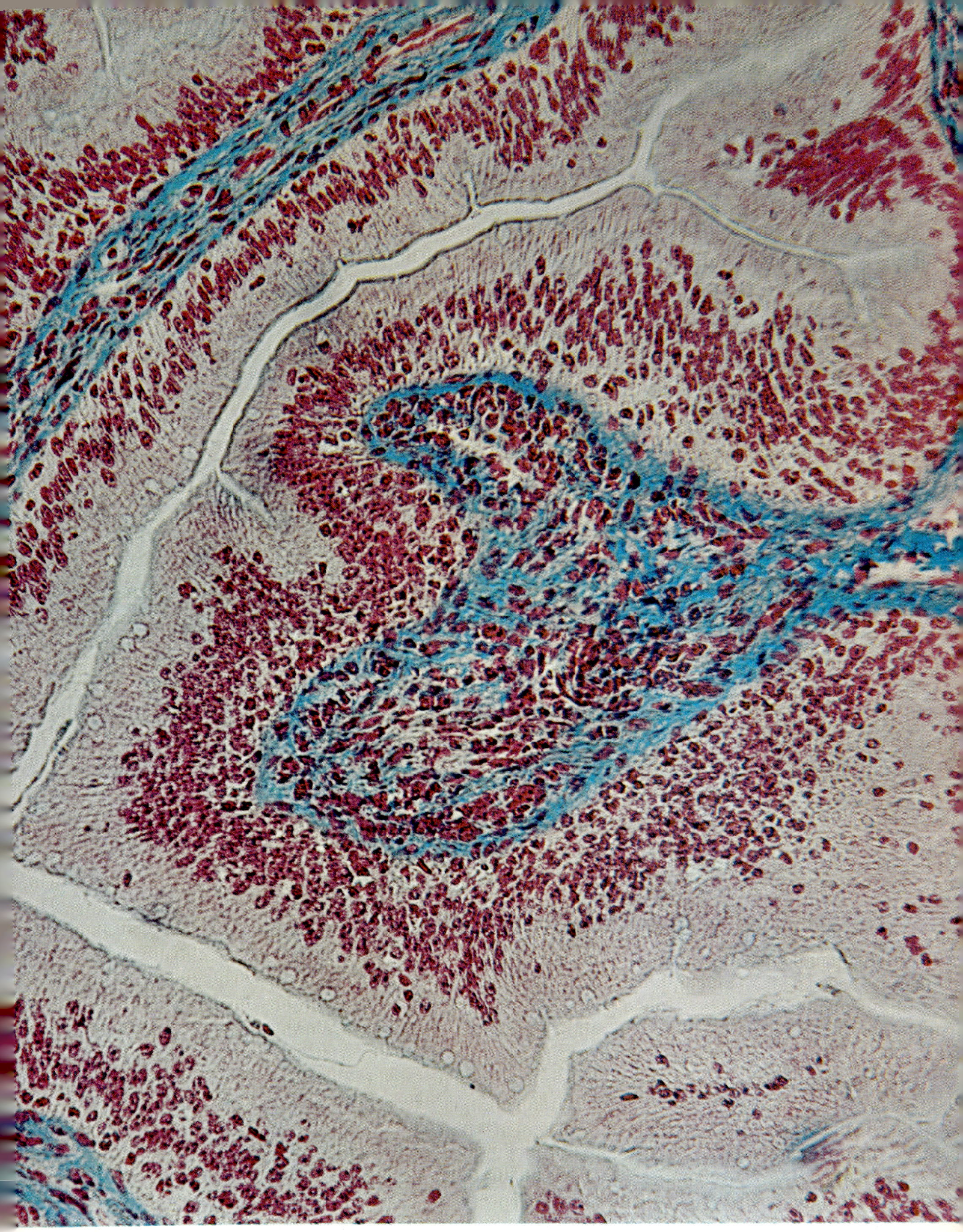

25 Small intestine of a frog. Magnification 600X.

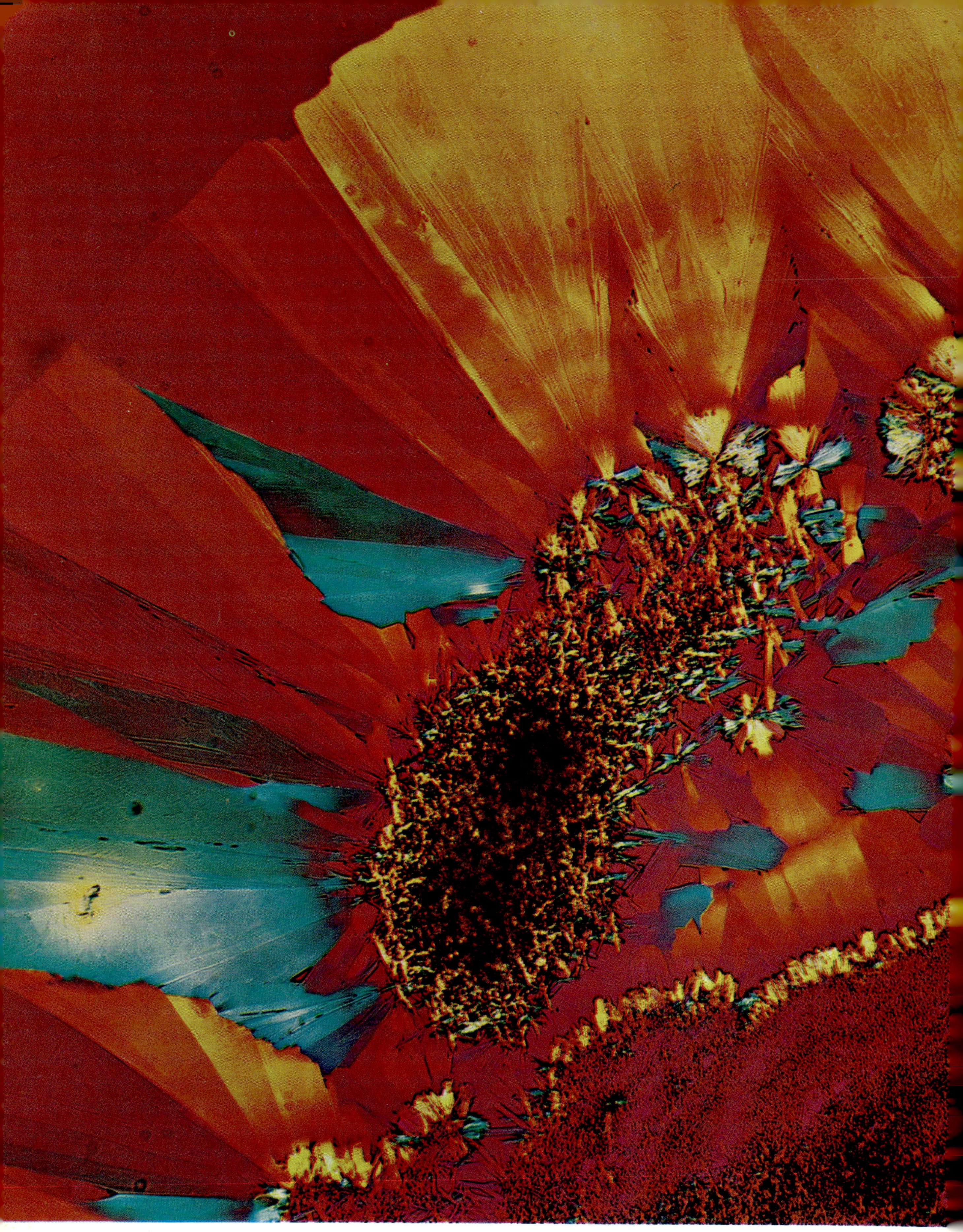

26 Sodium citrate. Magnification 160X.

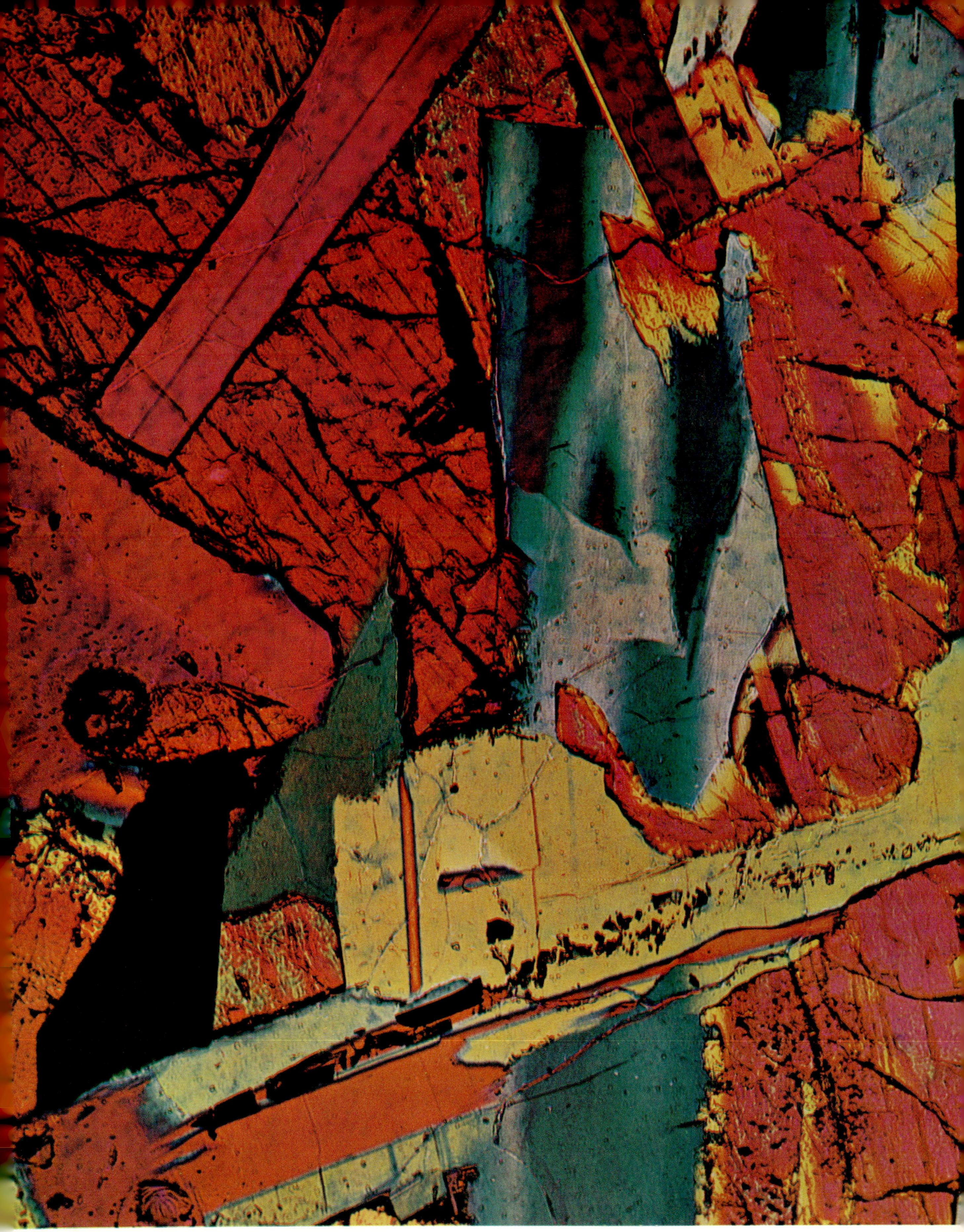

27 Moonrock. Magnification 400X.

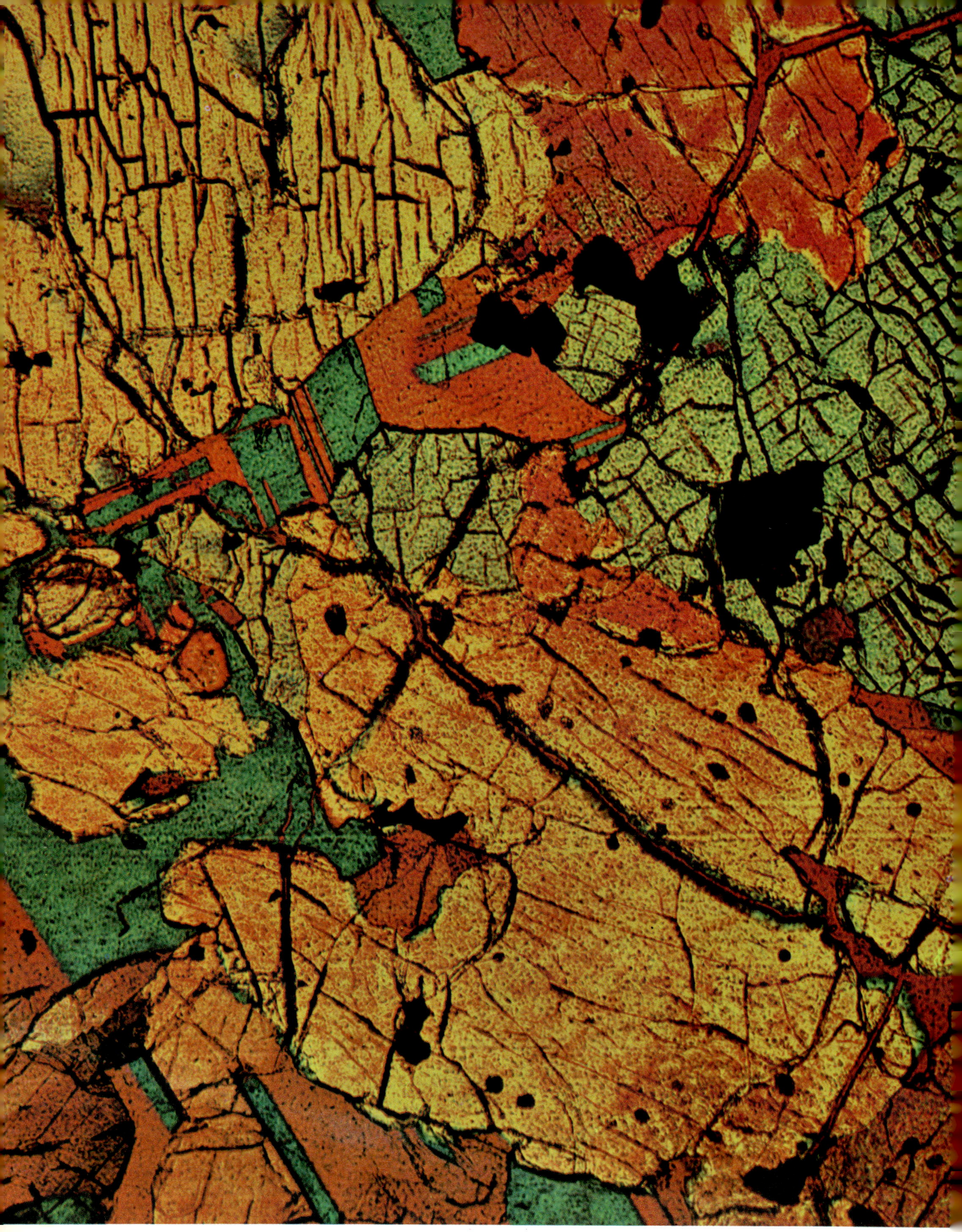

28 Moonrock. Magnification 400X.

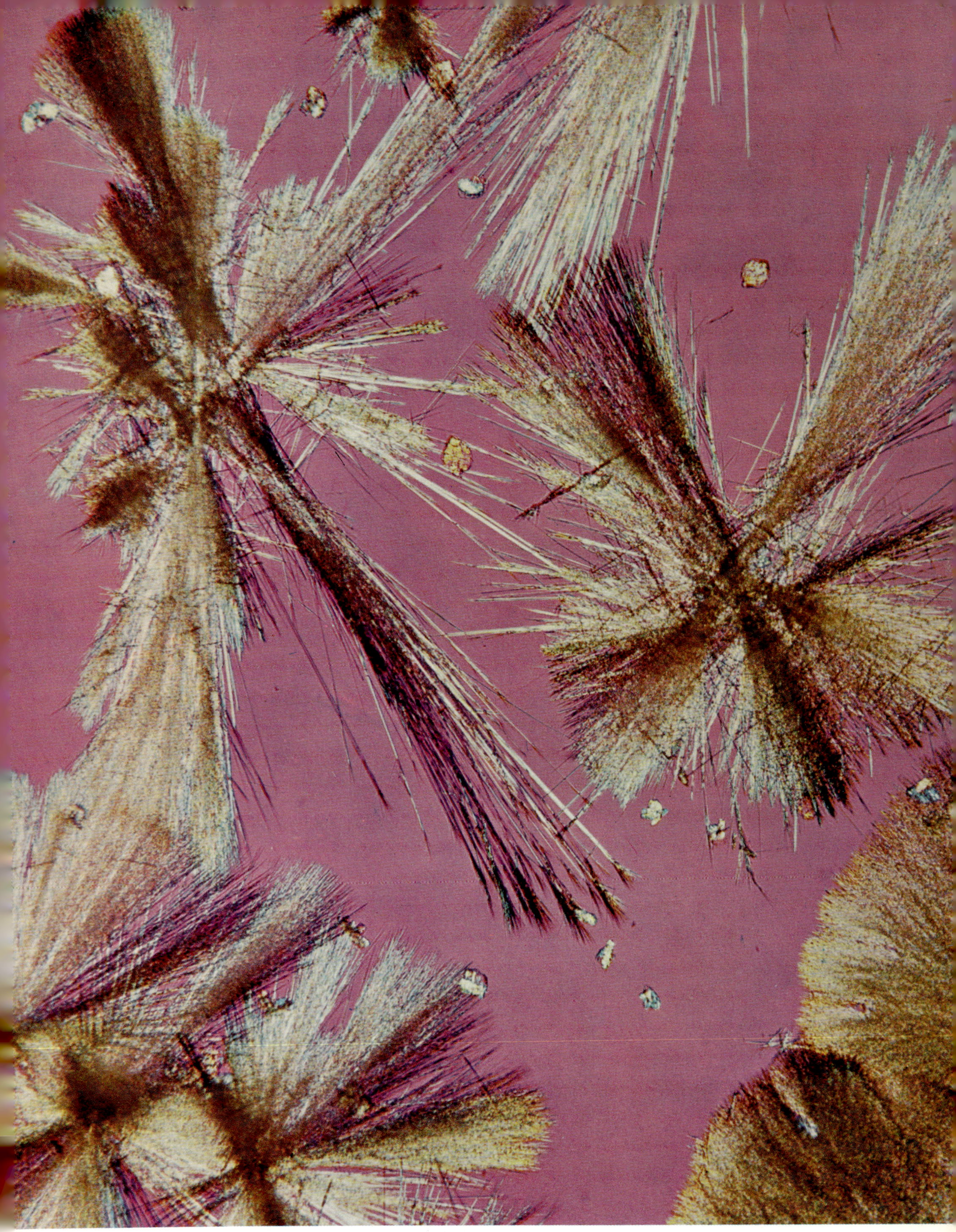

29 Quinidine Sulfate. Magnification 160X.

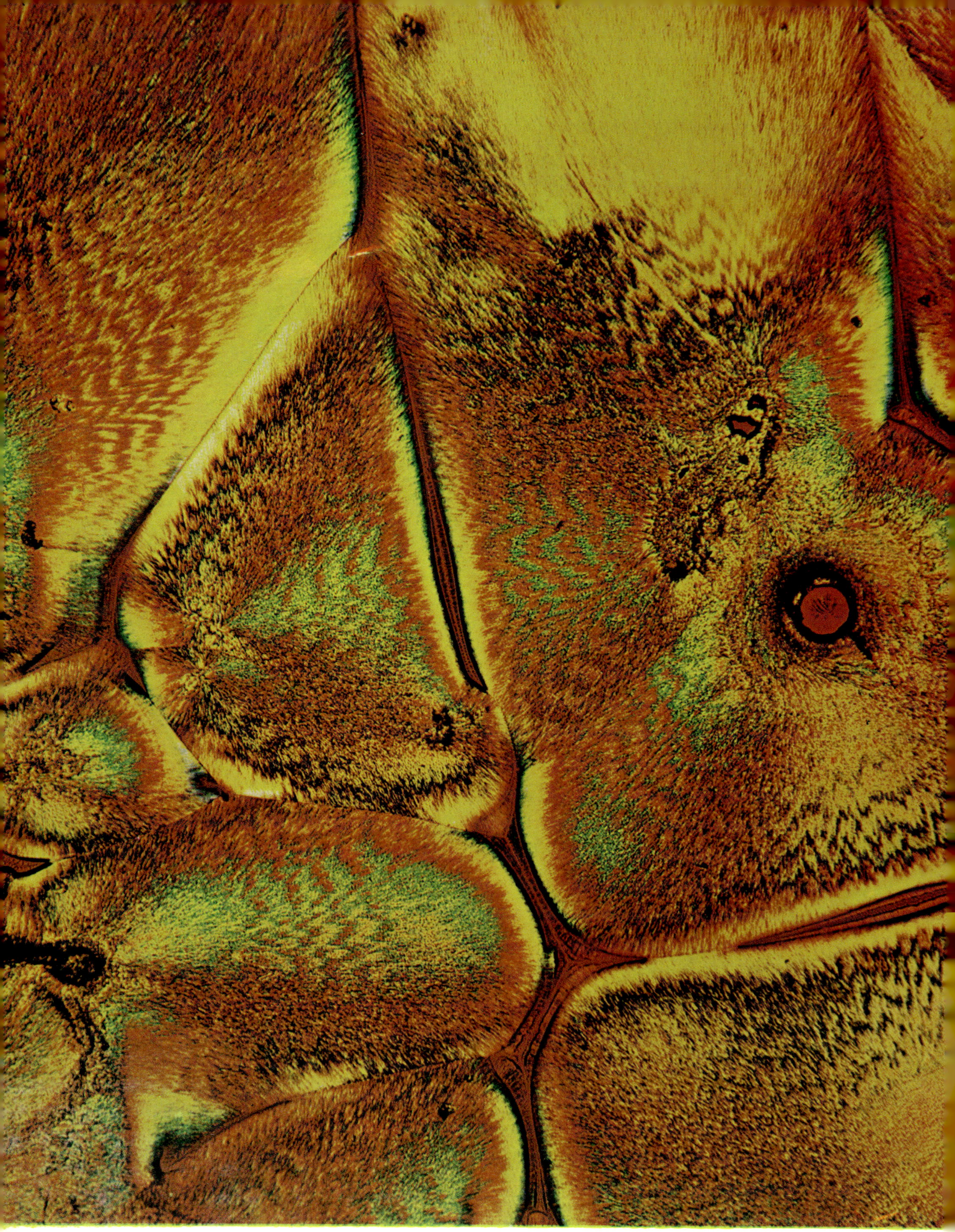

30 L-arginine. Magnification 400X.

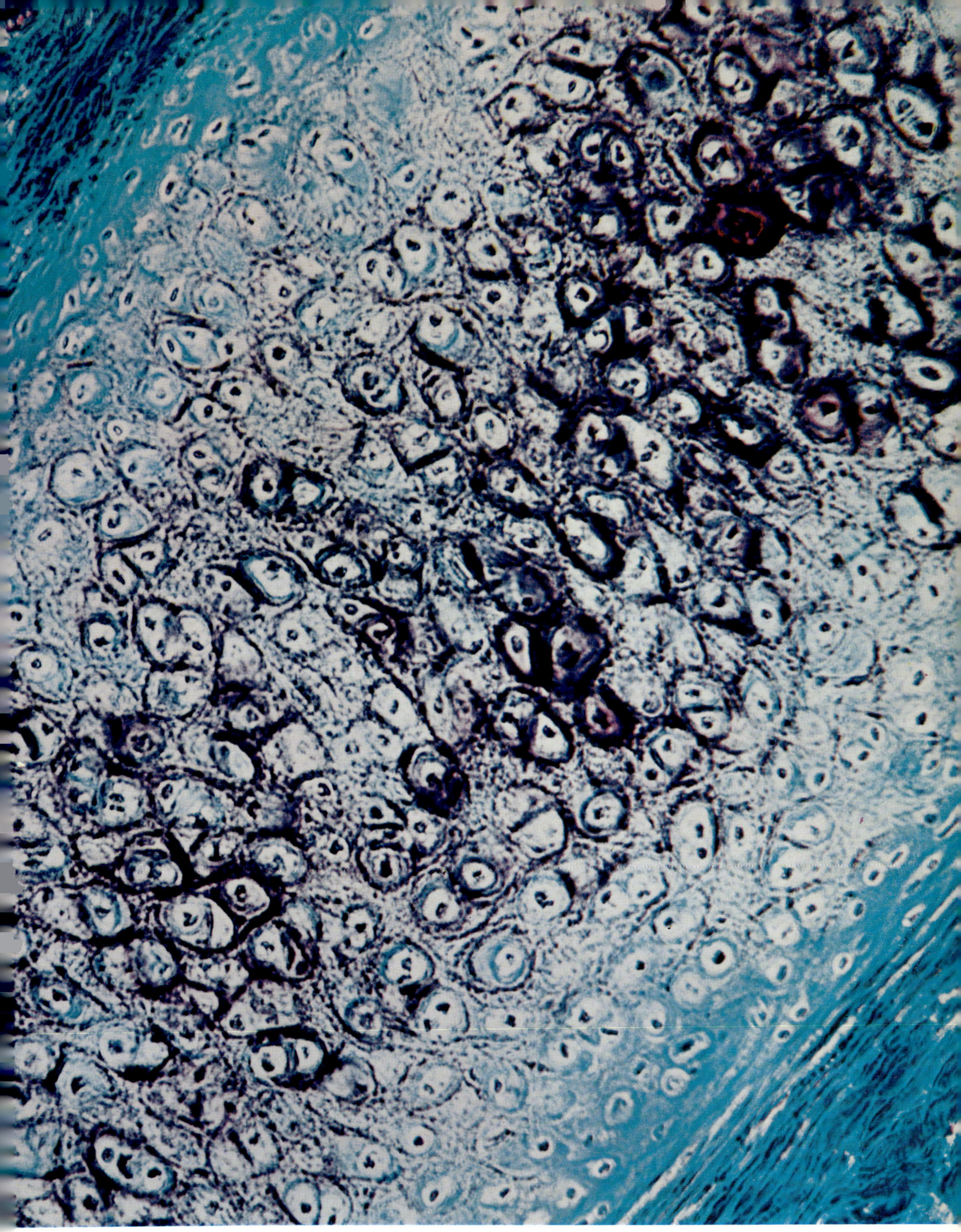

31 External ear. Magnification 800X.

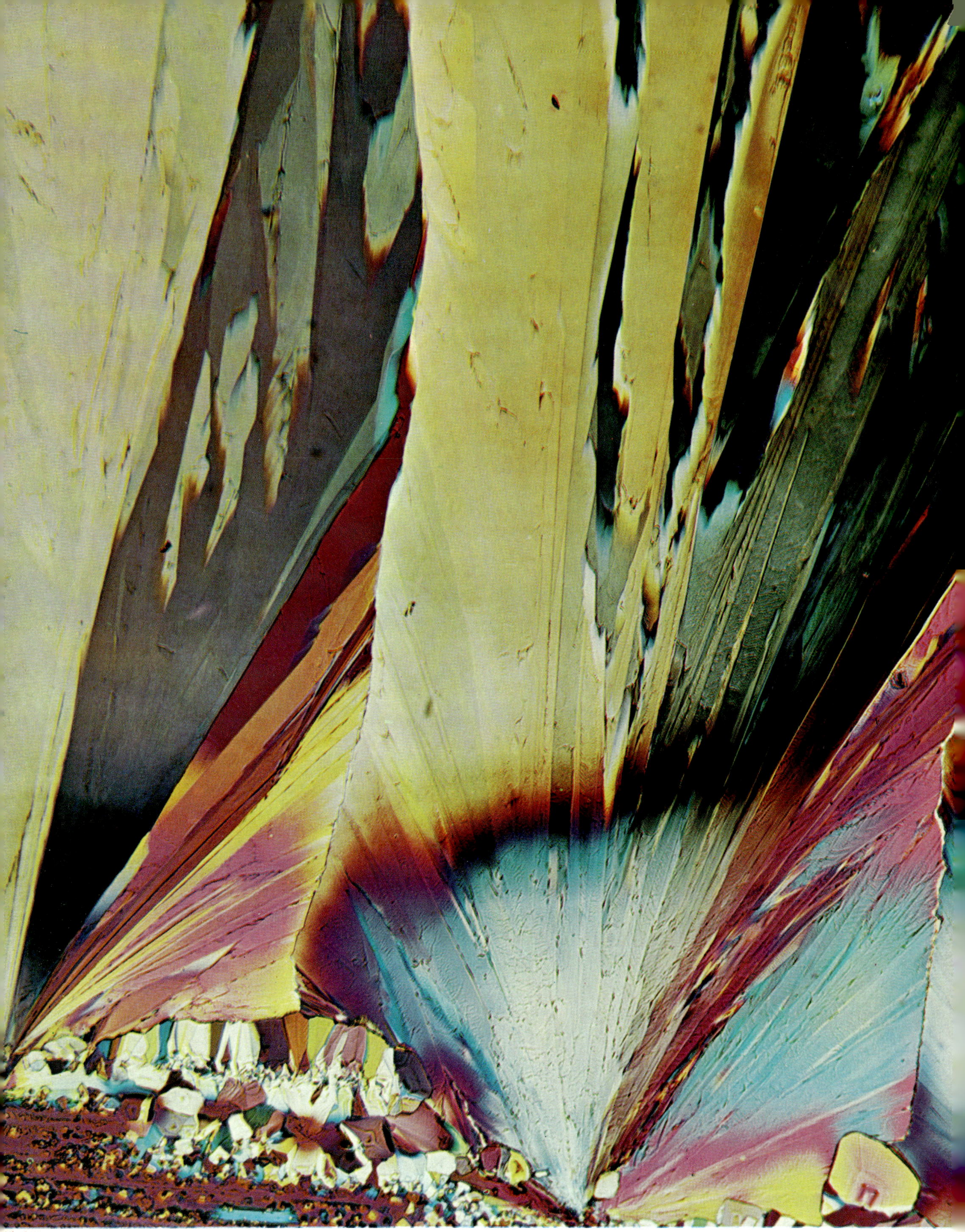

32 Sodium citrate. Magnification 160X.

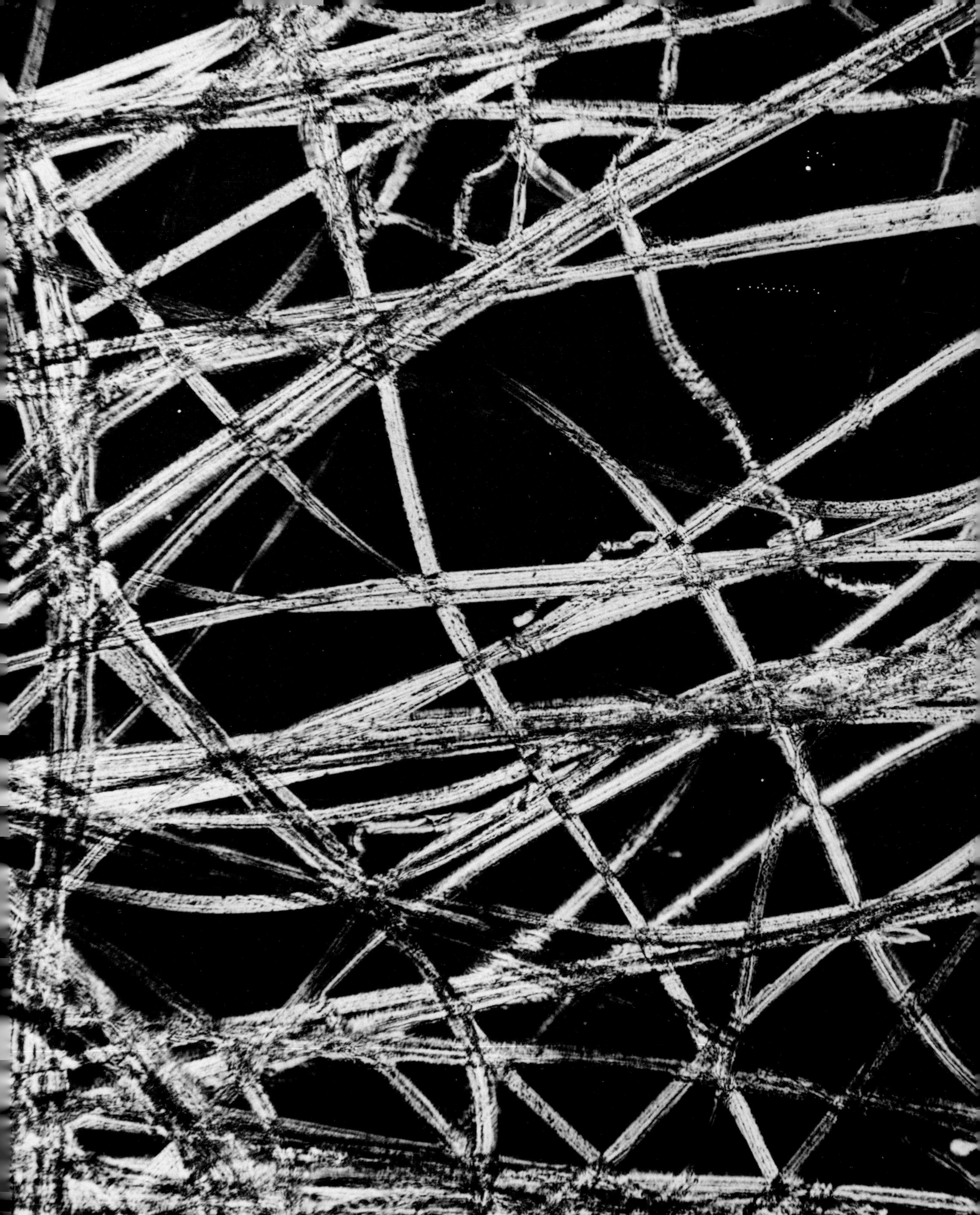

33 Pampers. Magnification 210X.

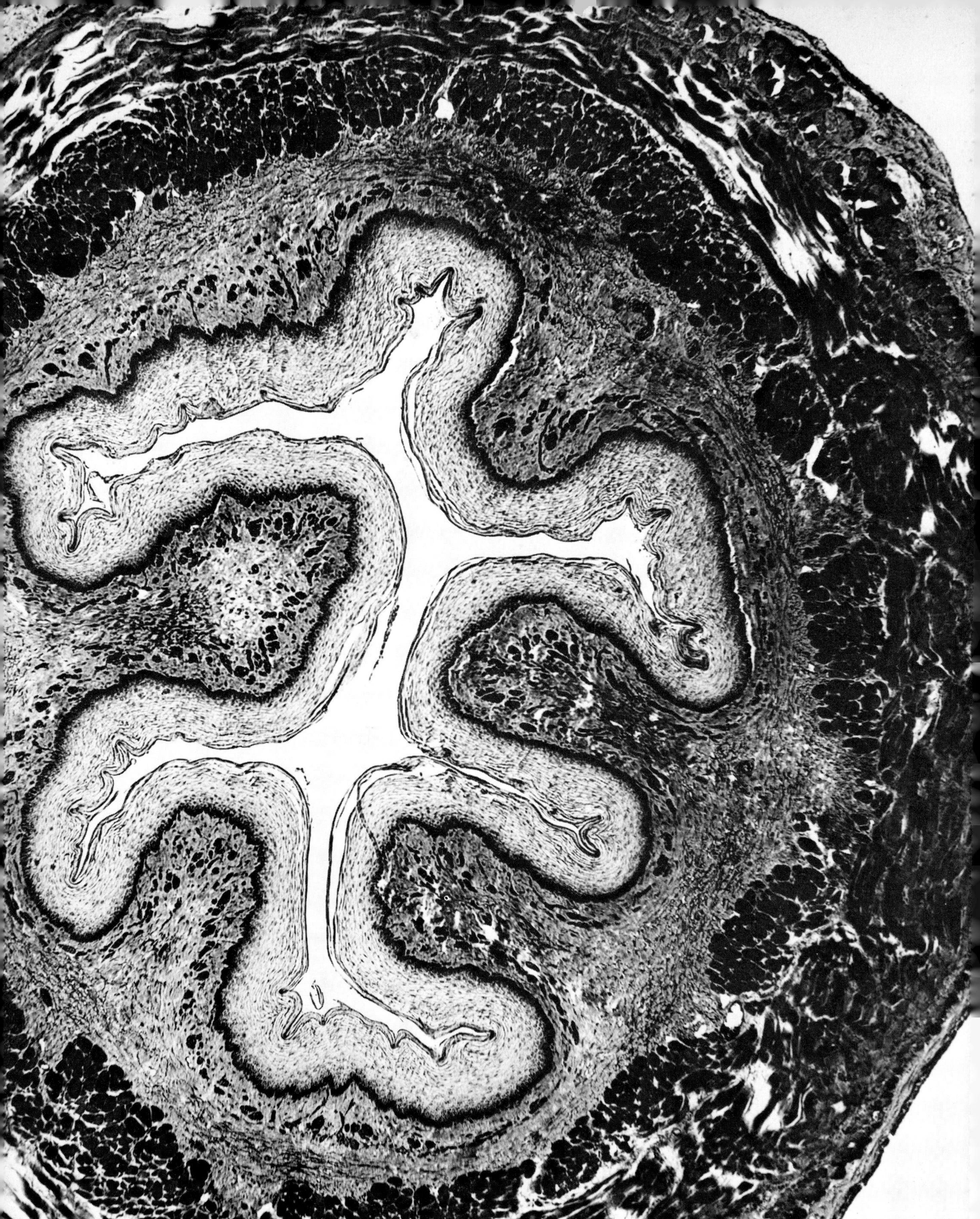

34 Rabbit esophagus. Magnification 190X.

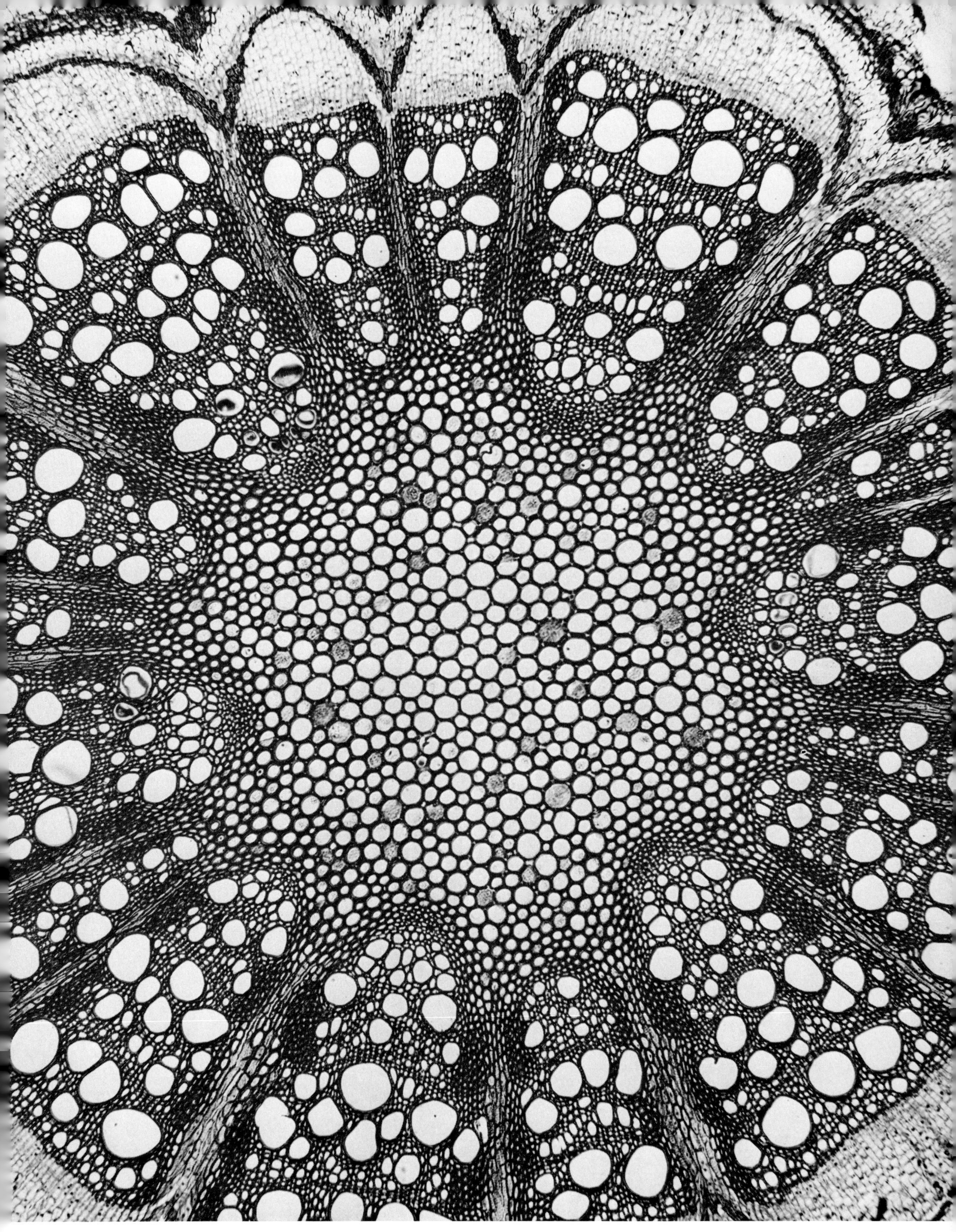

35 Clematis stem. Magnification 175X.

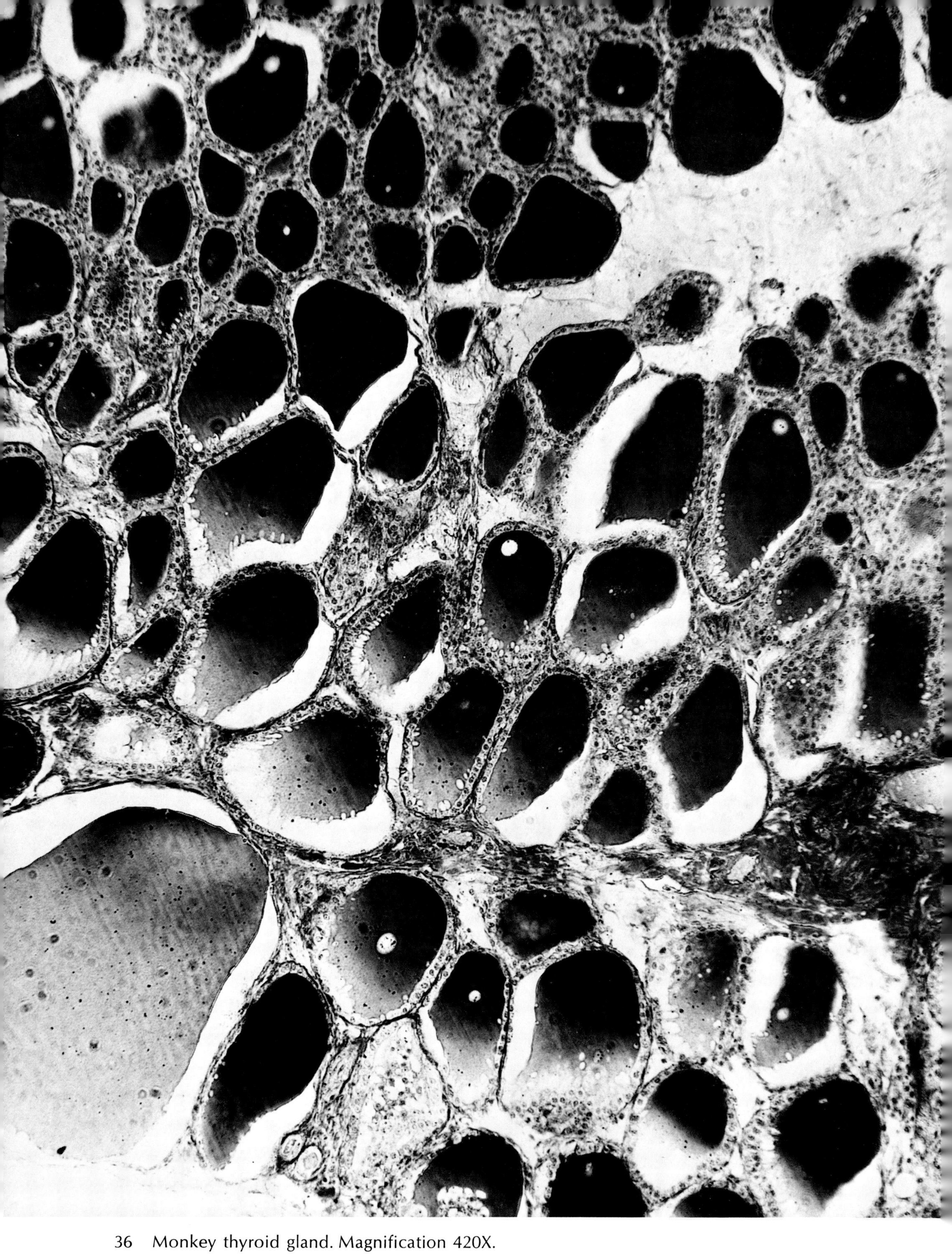

36 Monkey thyroid gland. Magnification 420X.

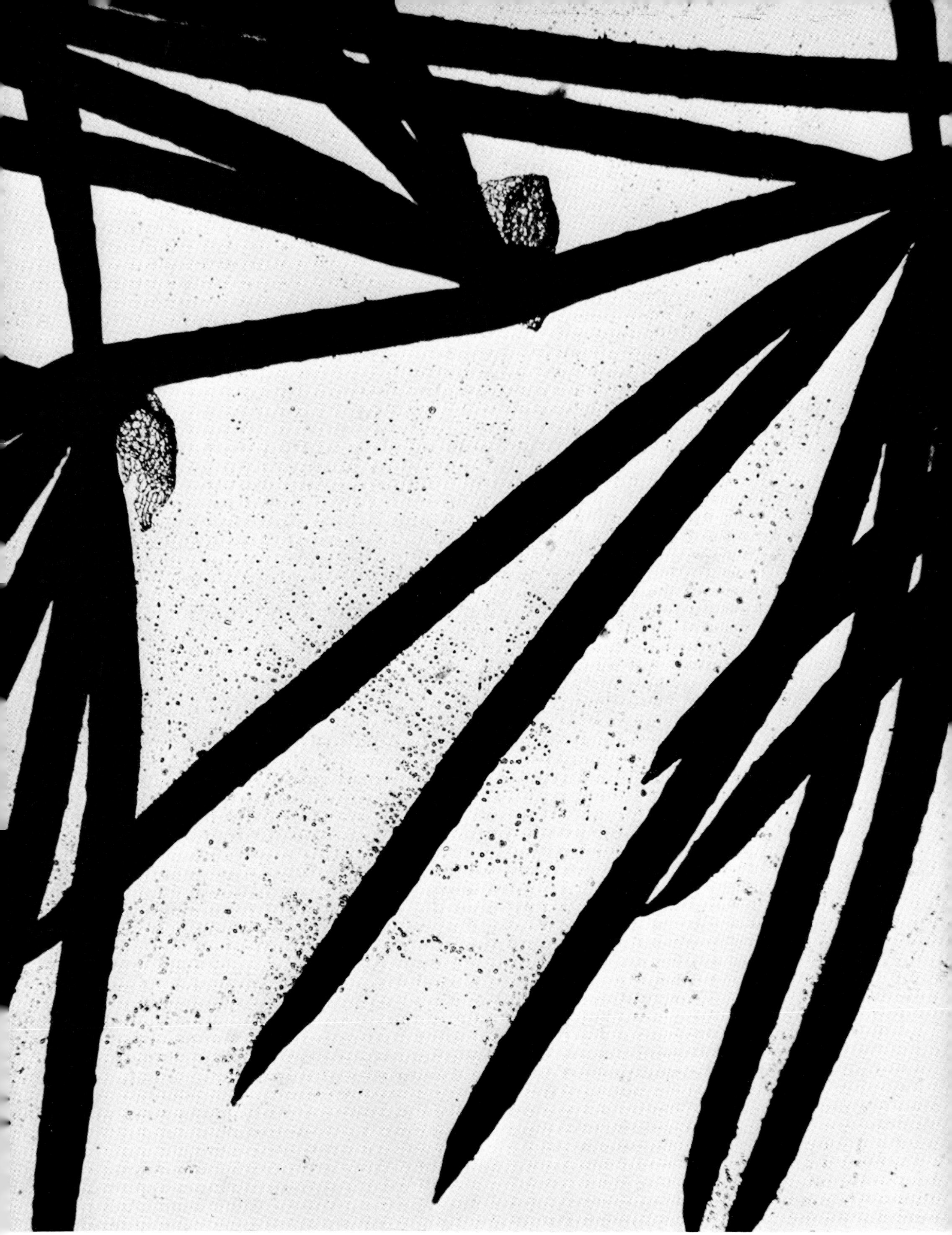

37 Fern. Magnification 140X.

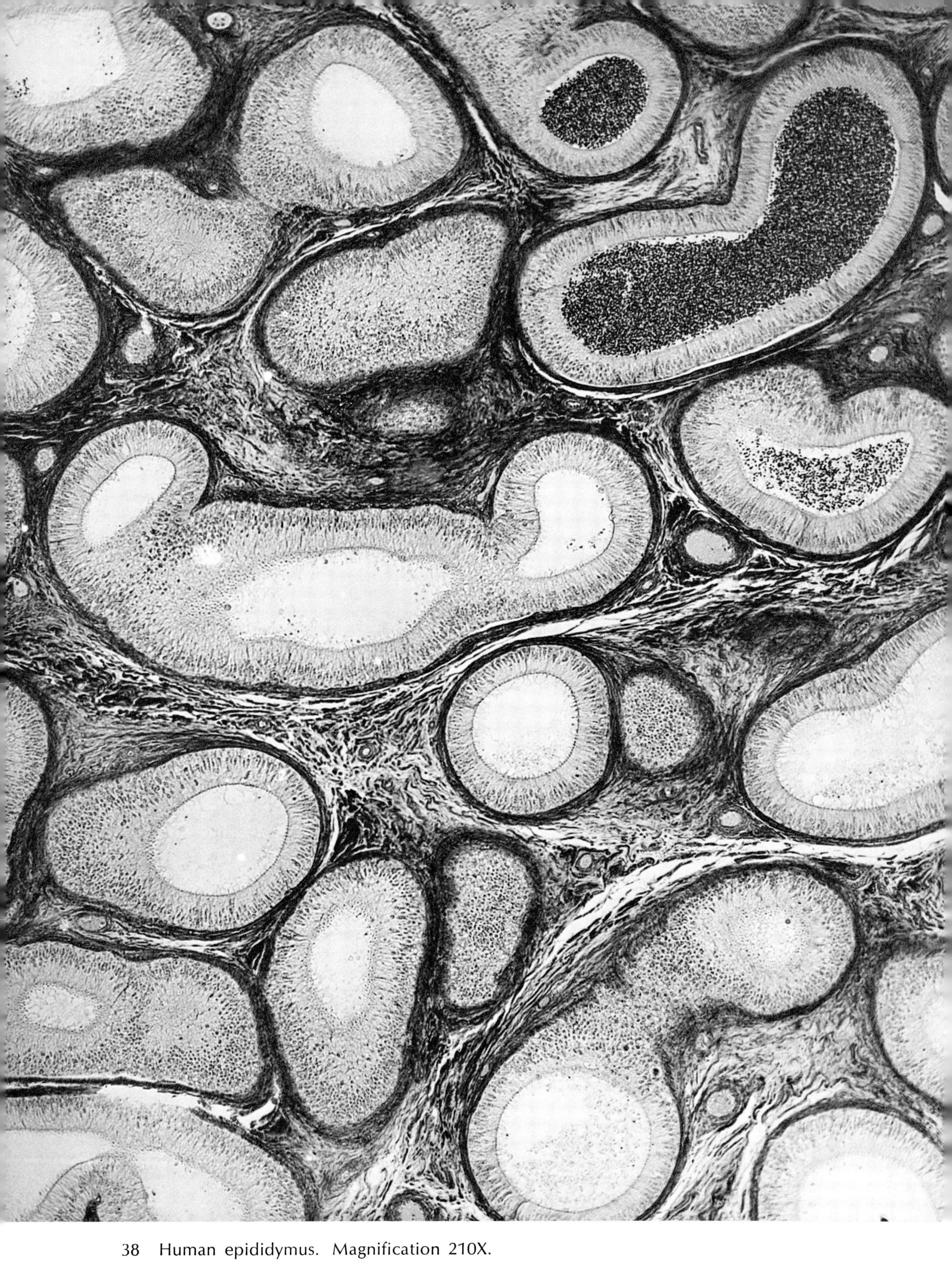

38 Human epididymus. Magnification 210X.

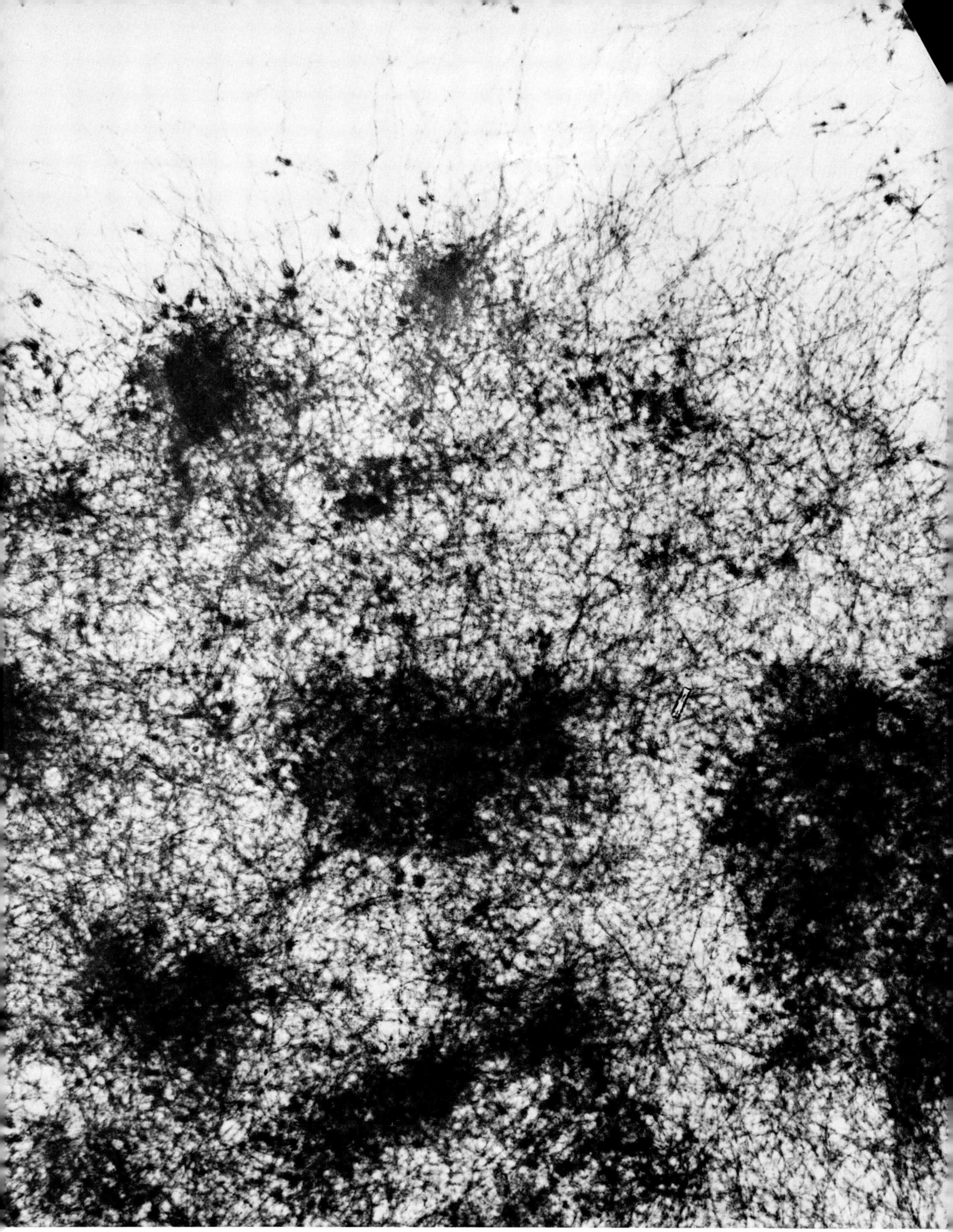

39 Penicillium. Magnification 455X.

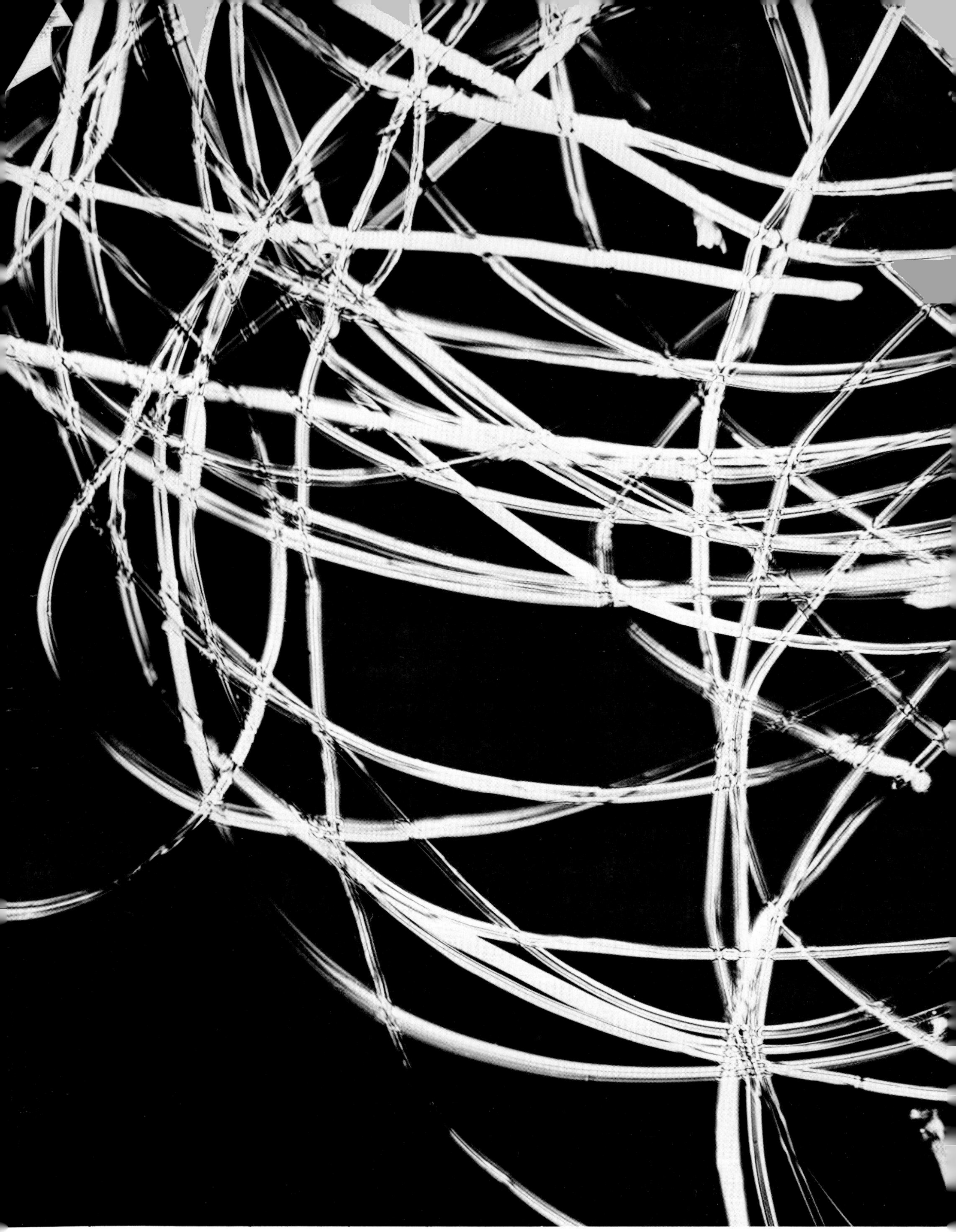

40 Silk. Magnification 350X.

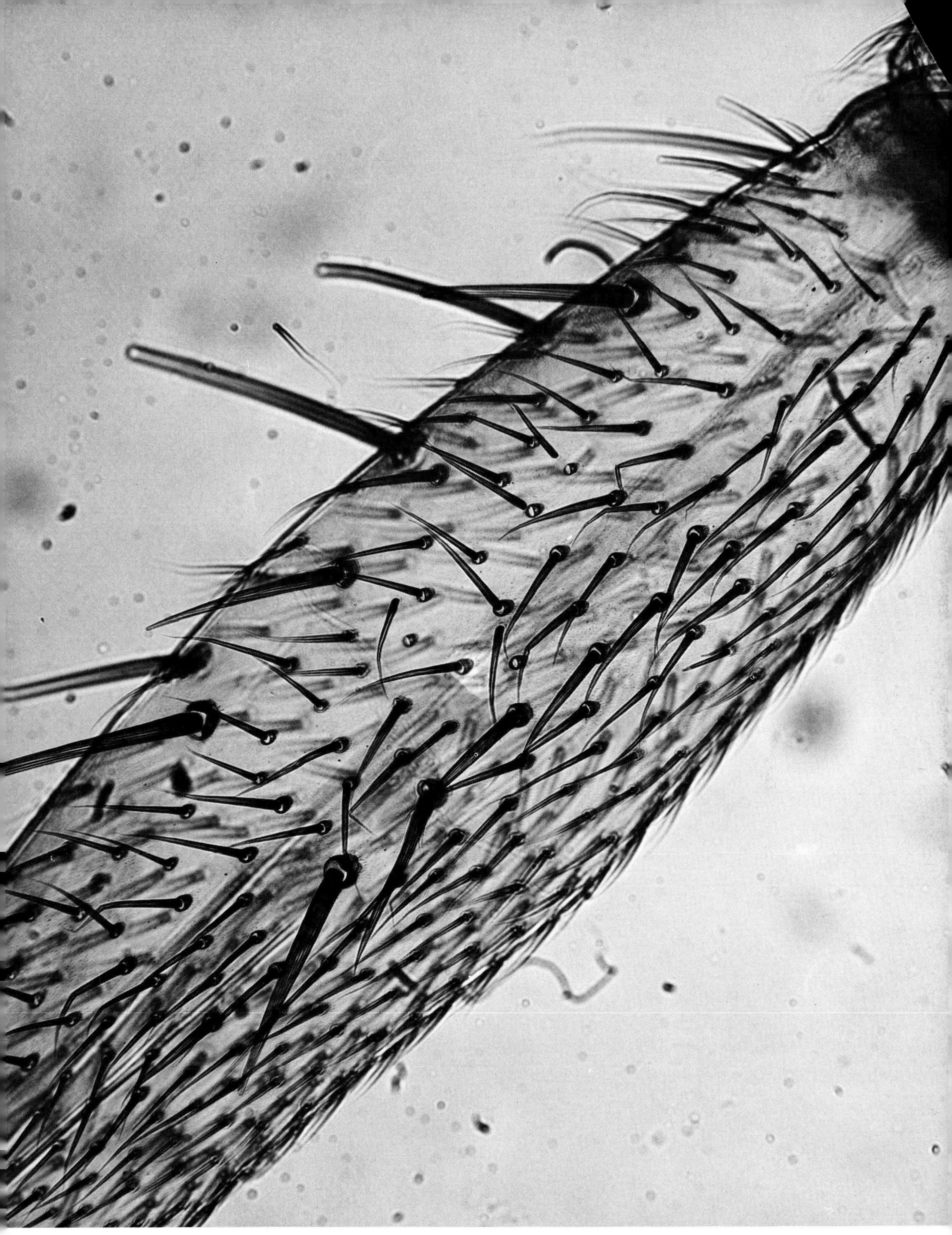

41 Housefly leg. Magnification 455X.

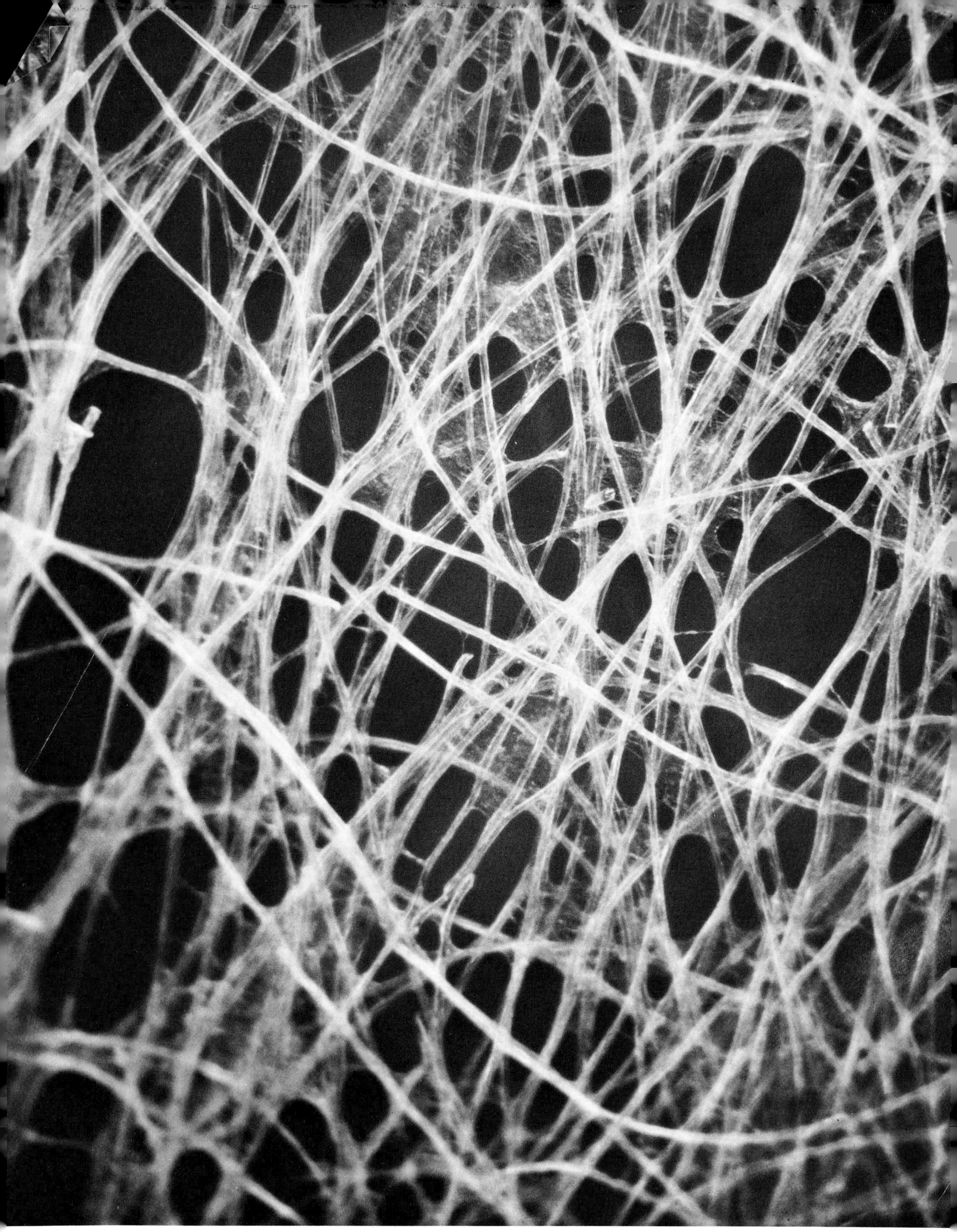

42 Lens paper. Magnification 175X.

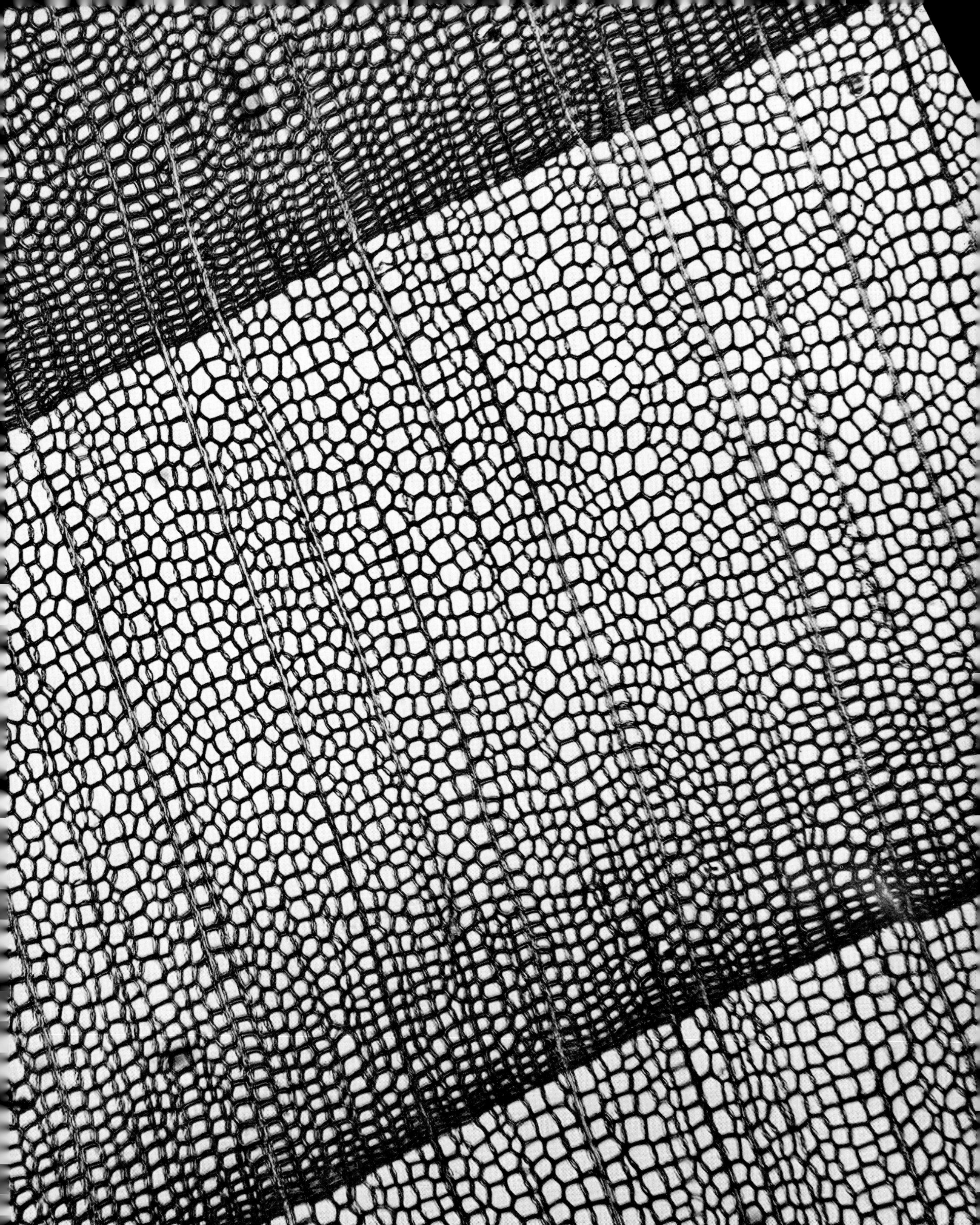

43 Pinus Strobus. Magnification 210X.

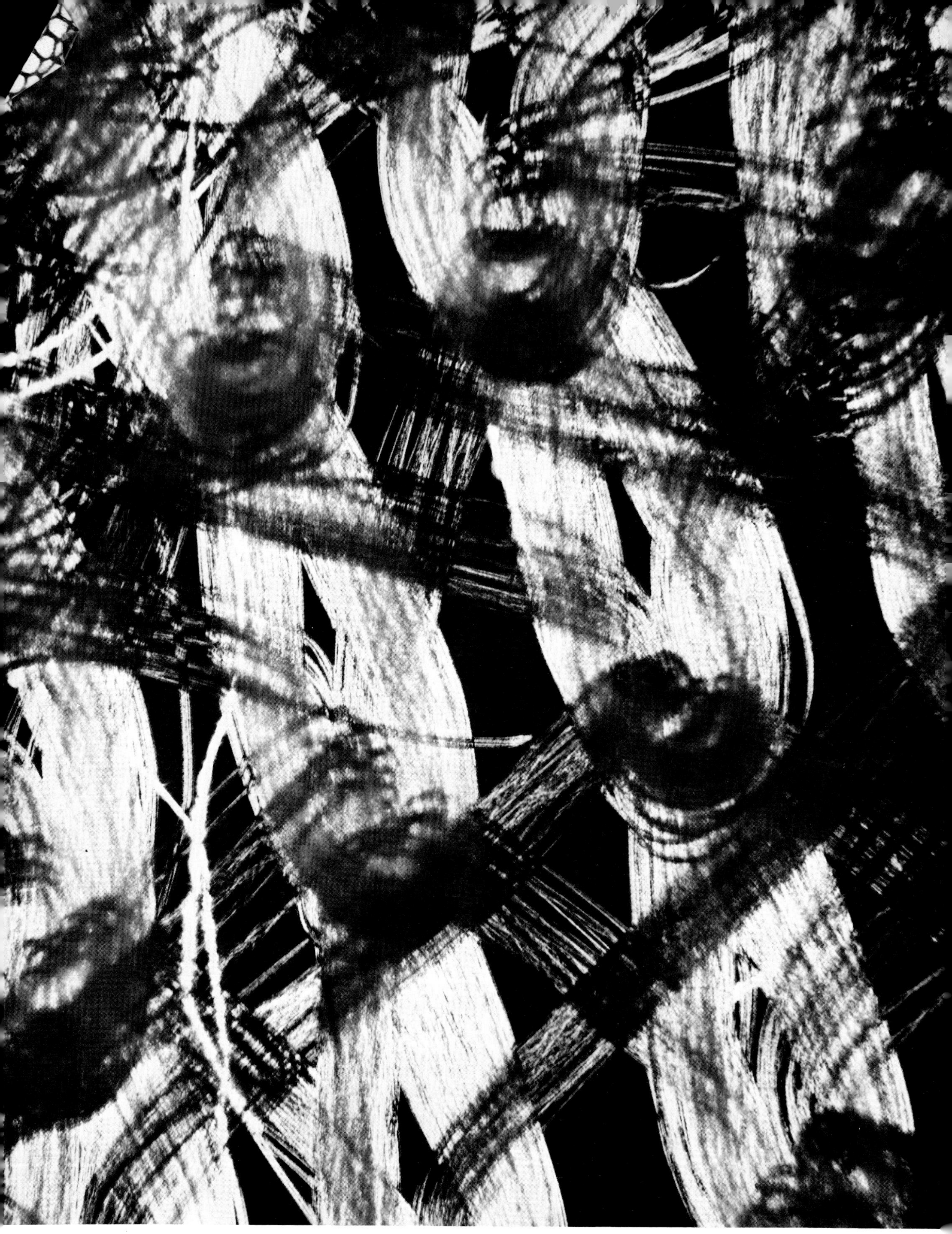

44 Cotton. Magnification 190X.

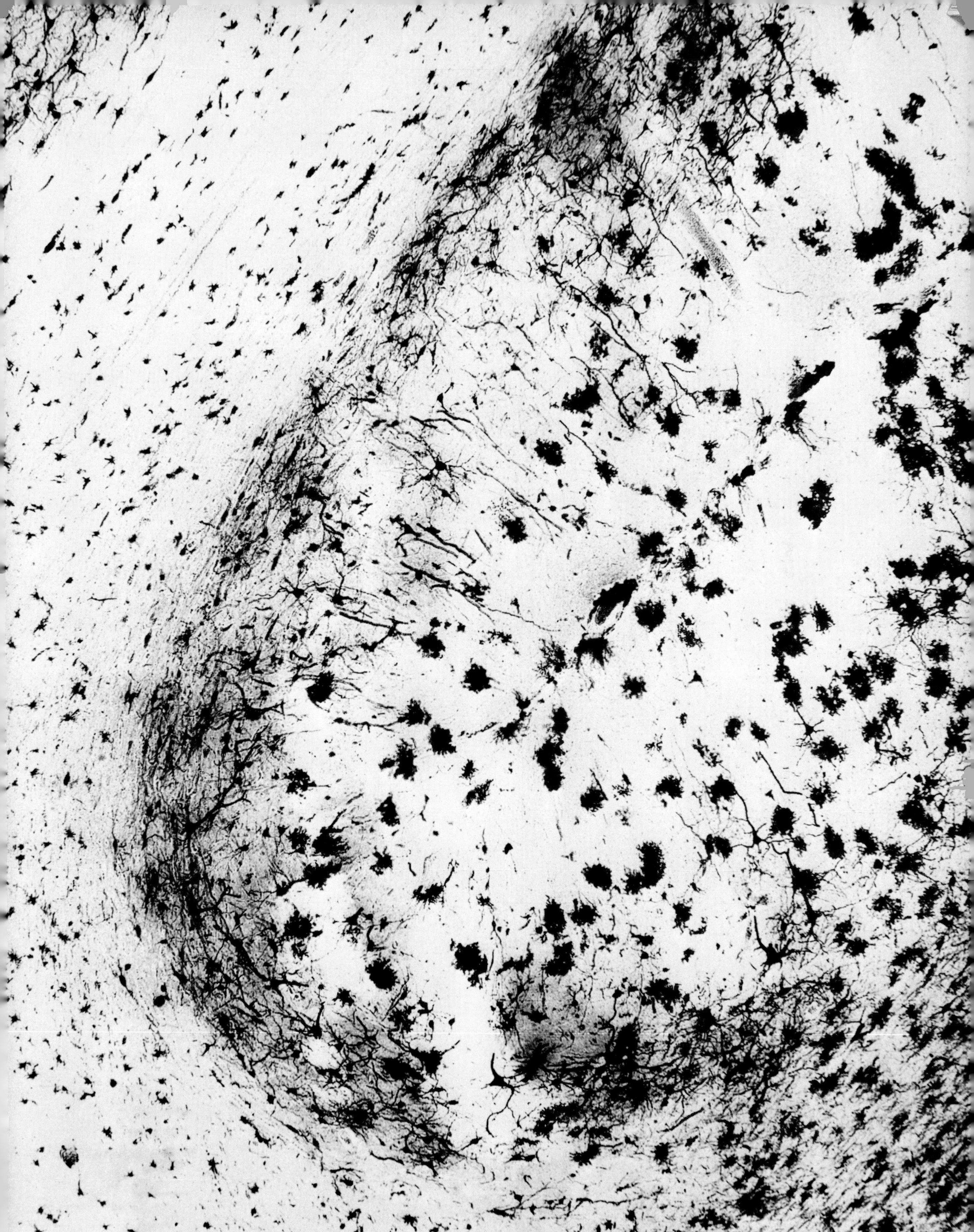

45 Cerebrum. Magnification 455X.

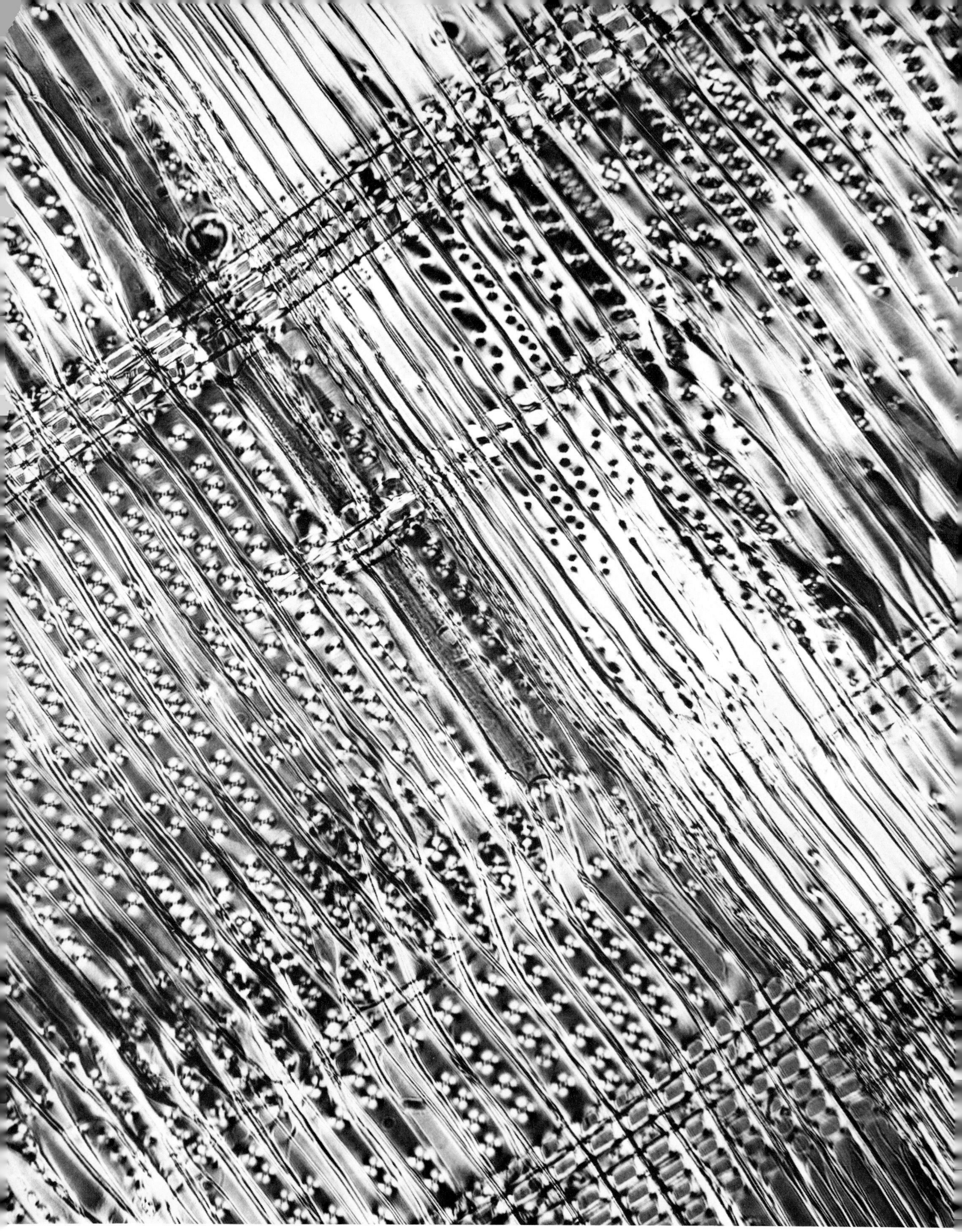

46 Pinus Strobus. Magnification 420X.

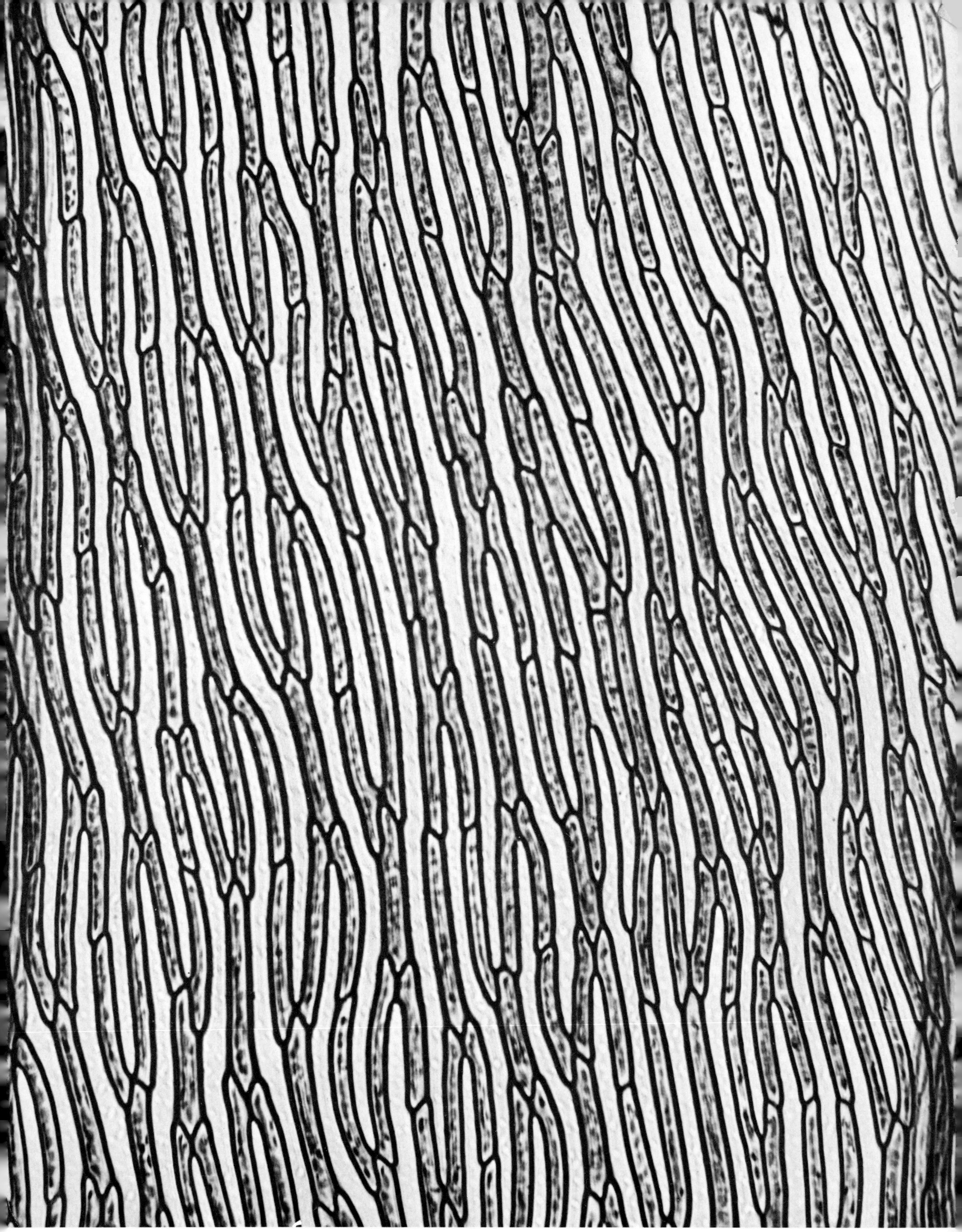

47 Sphagnum leaf (bog moss). Magnification 700X.

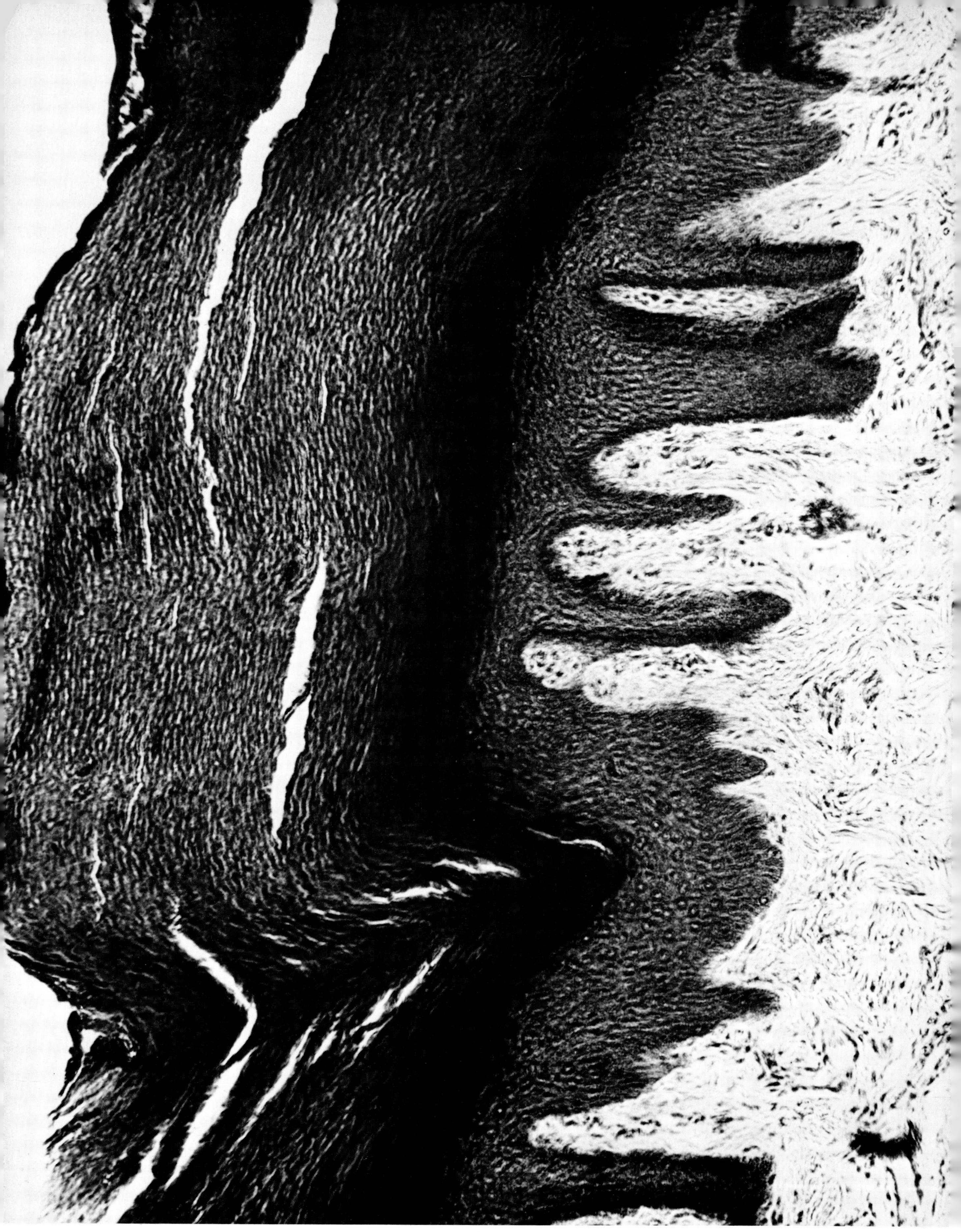

48 Human finger tip. Magnification 455X.

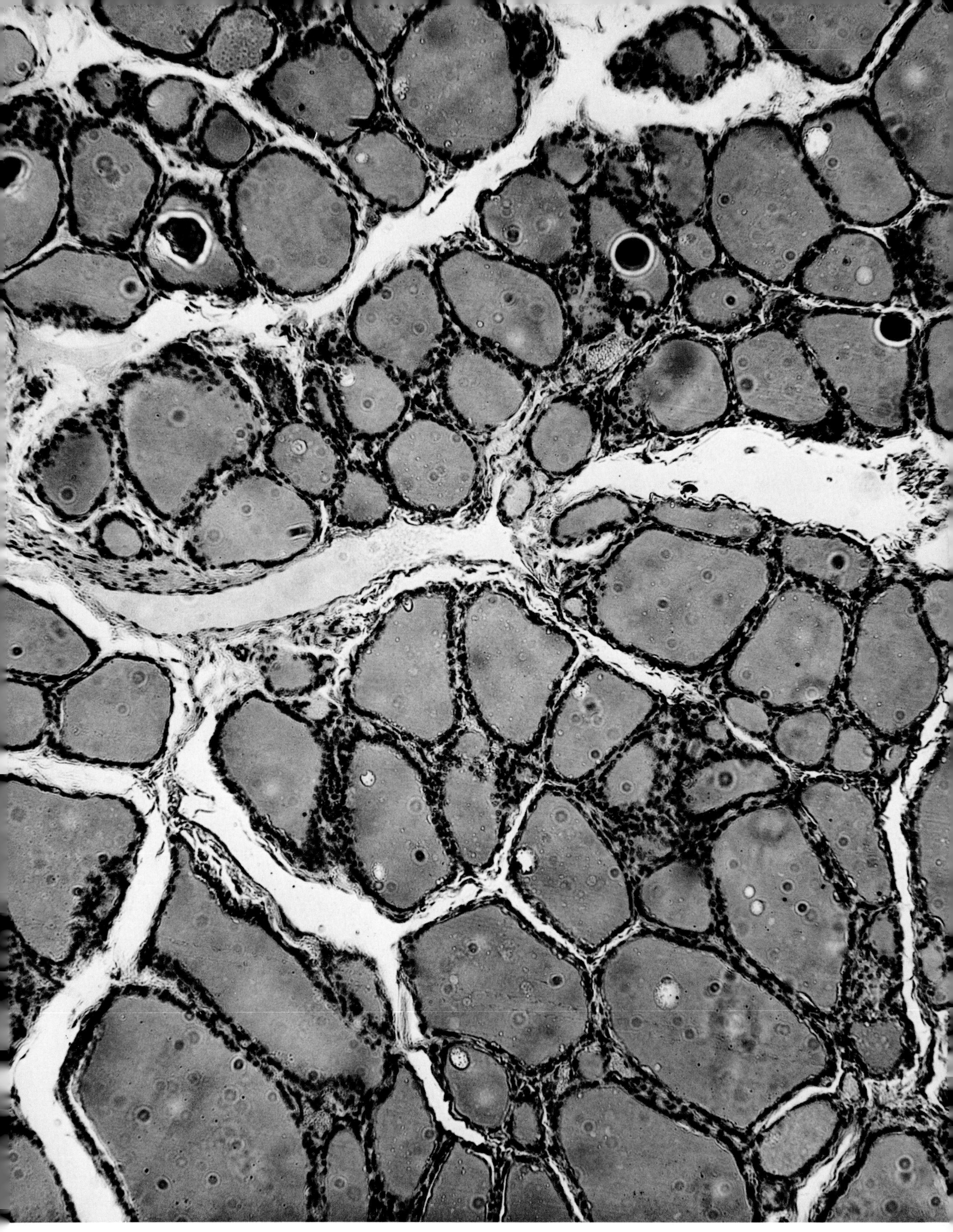

49 Human thyroid. Magnification 420X.

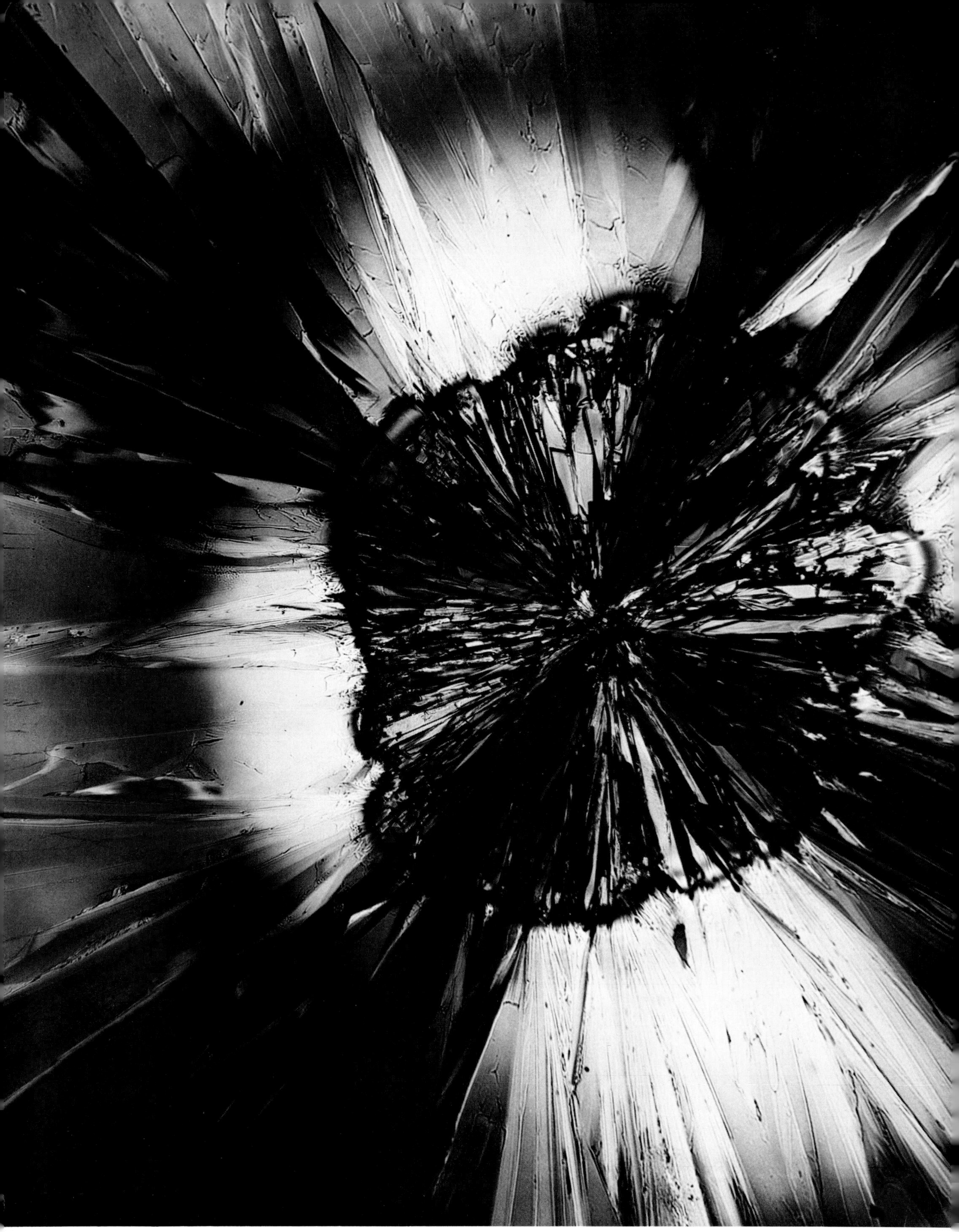

50 Sodium citrate. Magnification 210X.

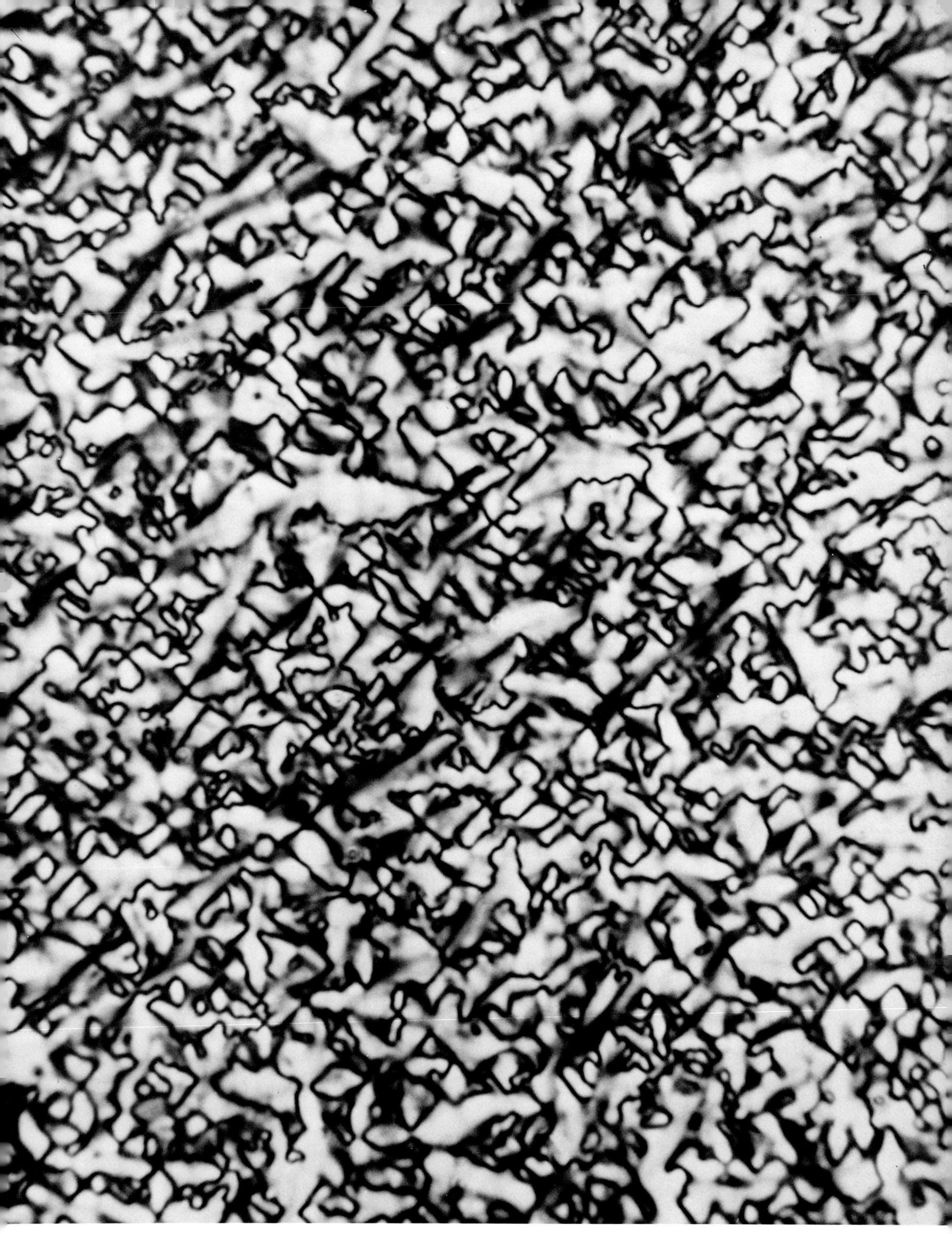

51 Ester of cholesterol. Magnification 2100X.

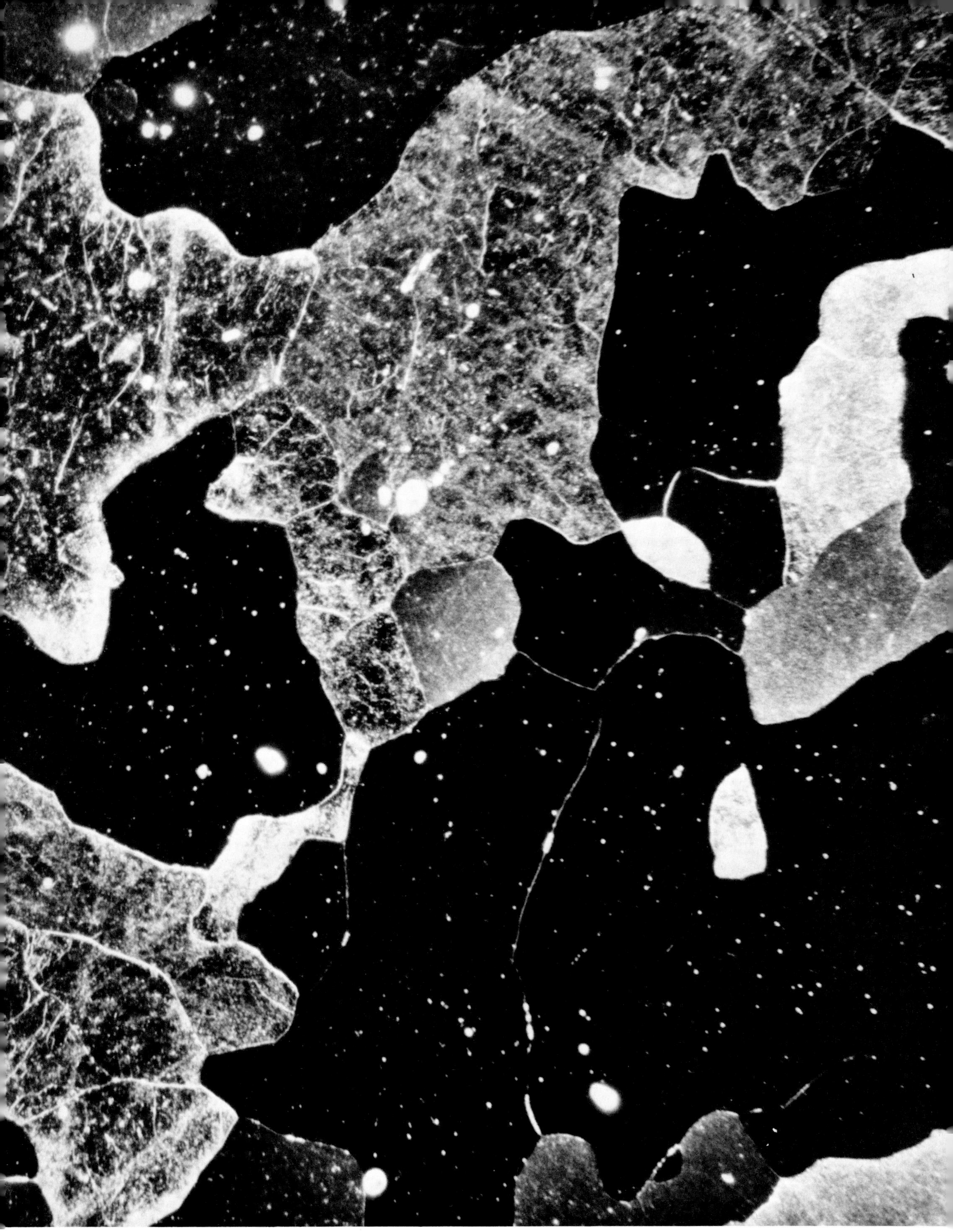

52 Carbon iron alloy. Magnification 225X.

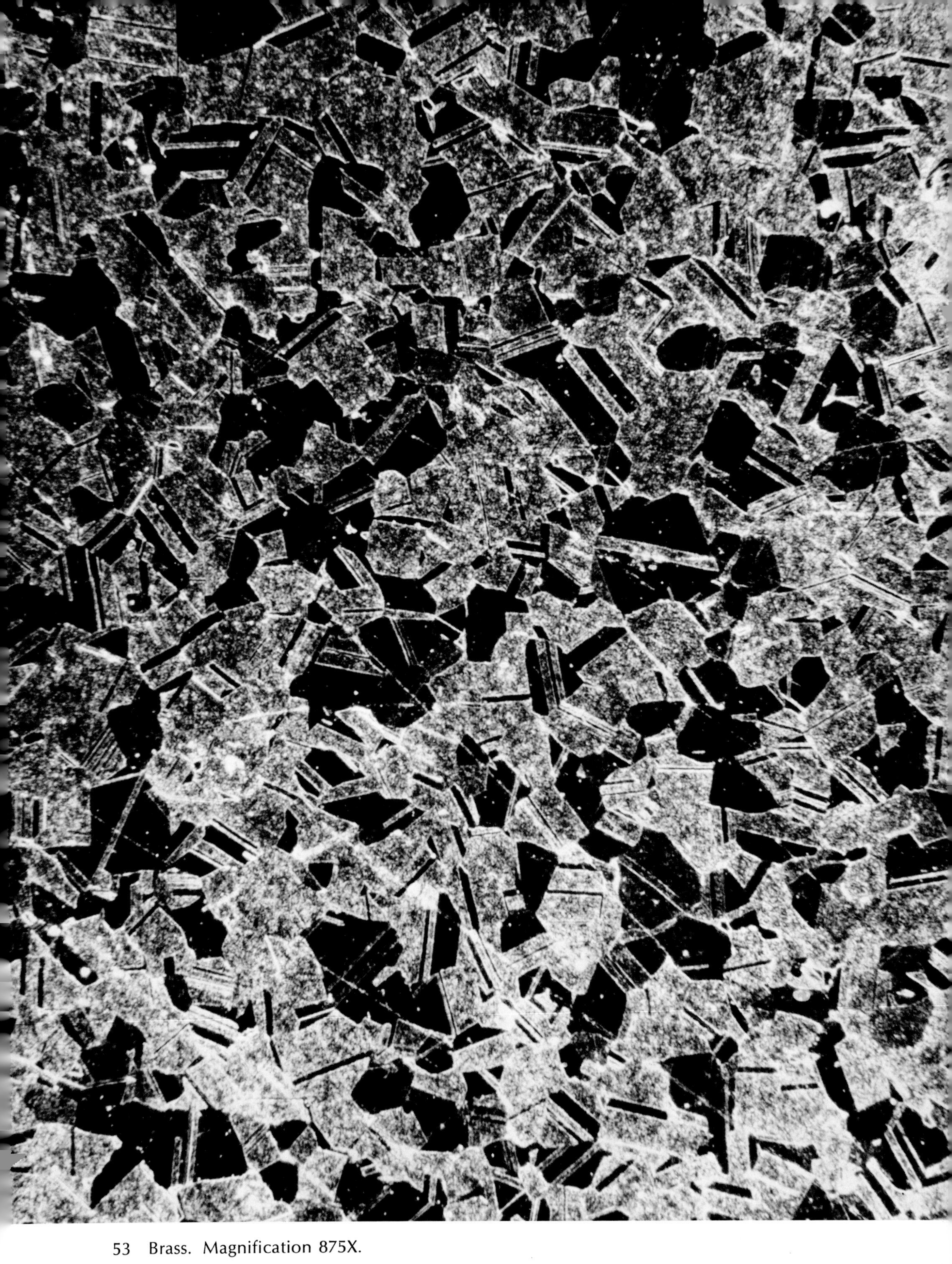

53 Brass. Magnification 875X.

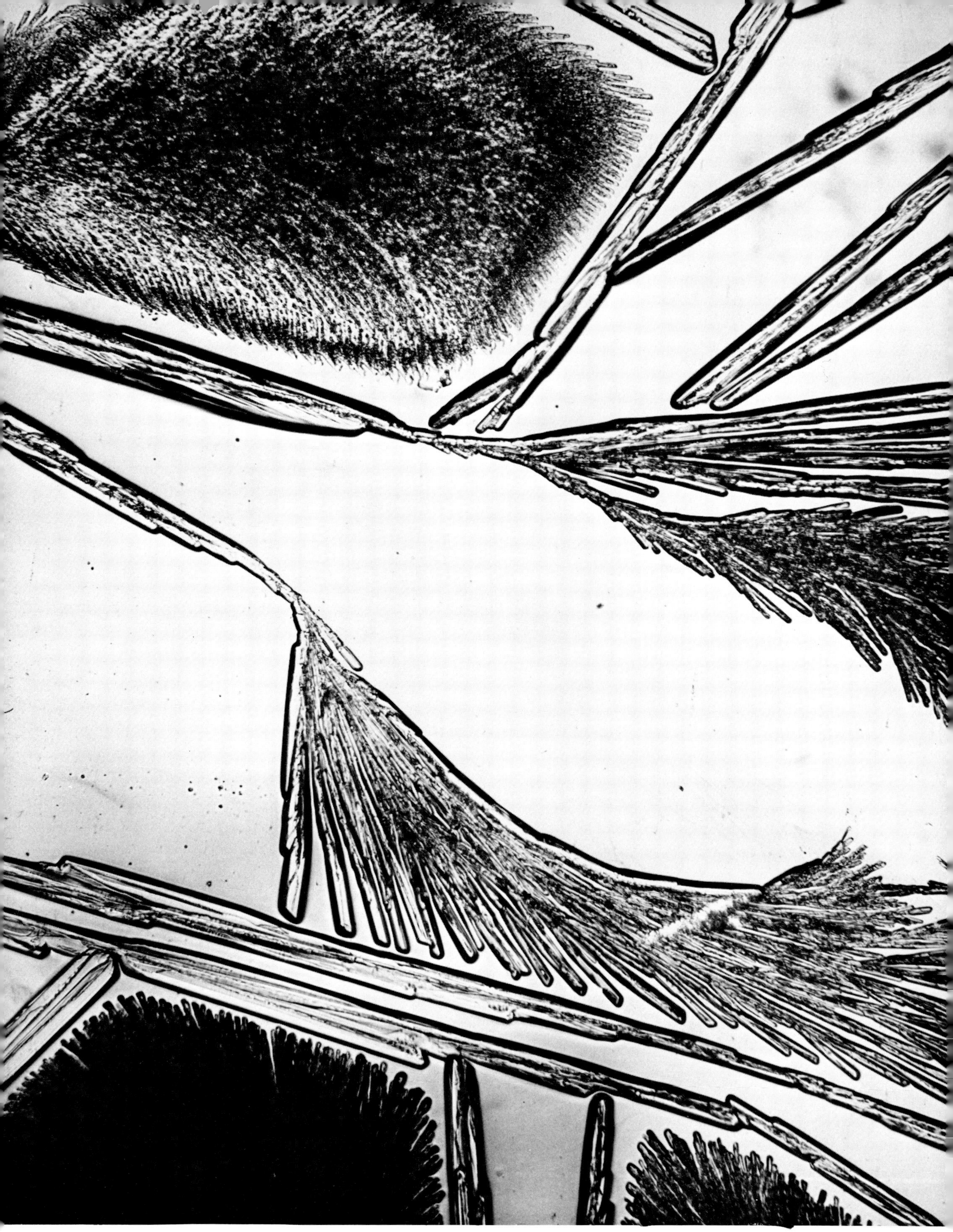

54 Camphor and cyclohexone. Magnification 225X.

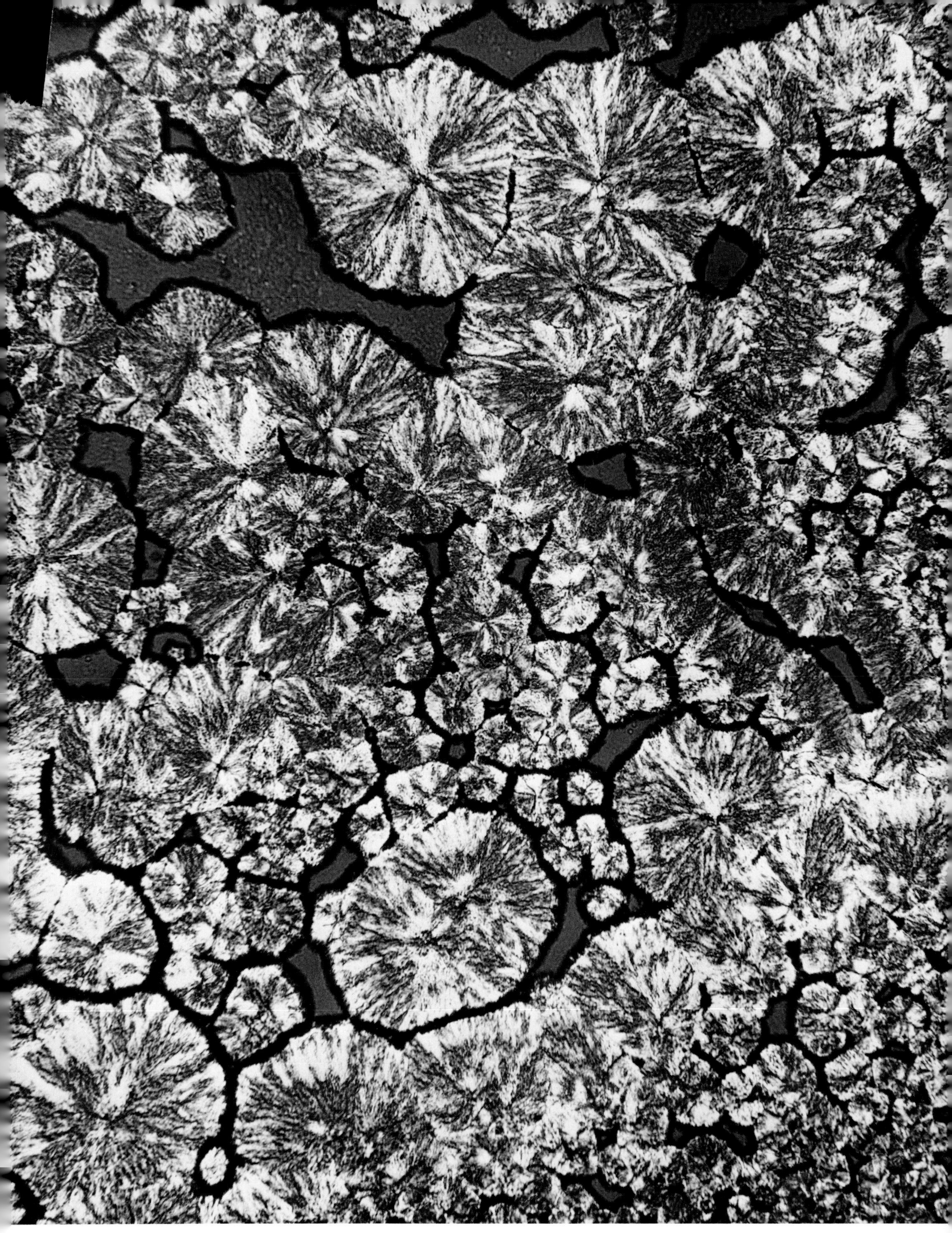

55 D-lysine. Magnification 875X.

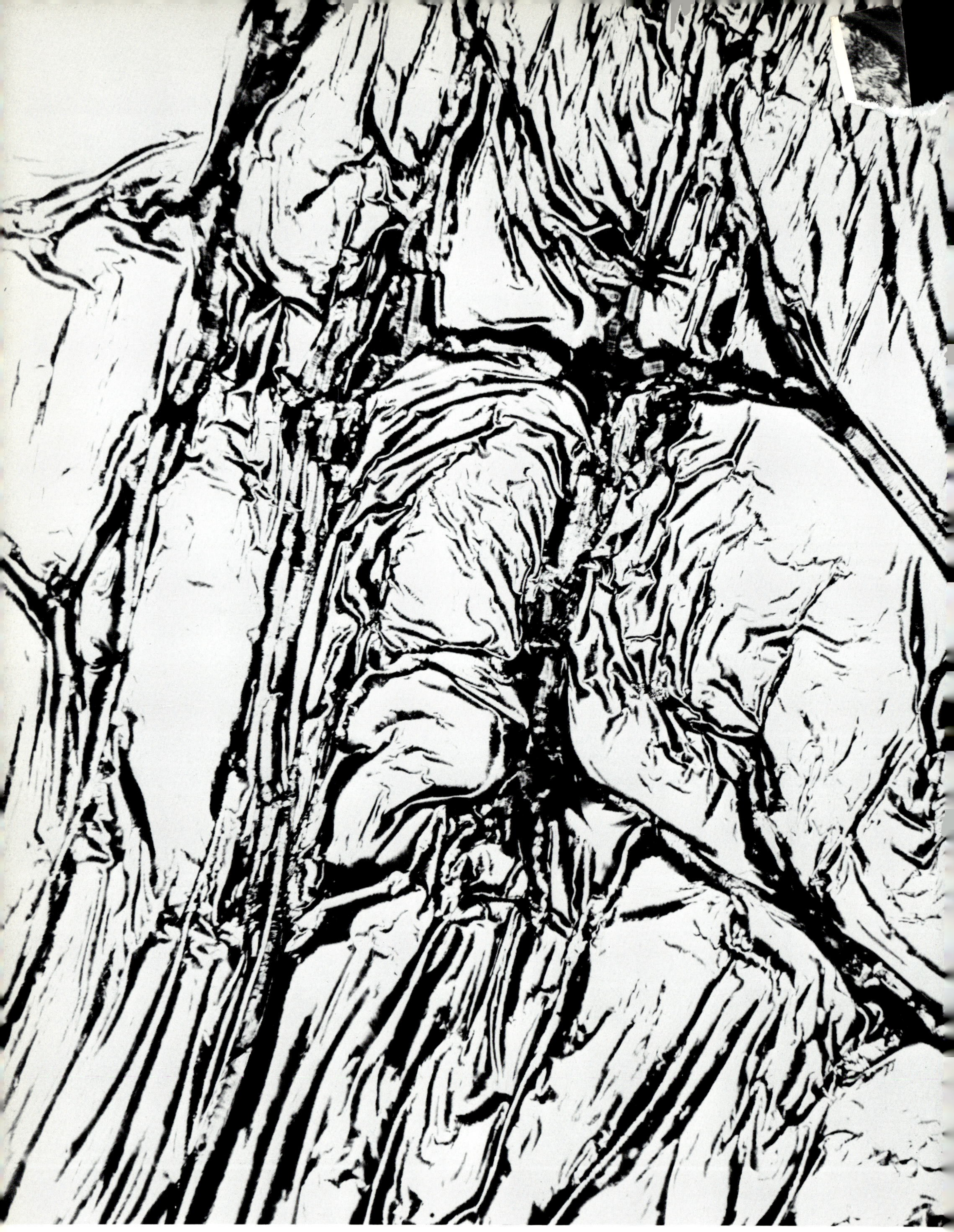

56 Ferric ammonium sulfate. Magnification 225X.

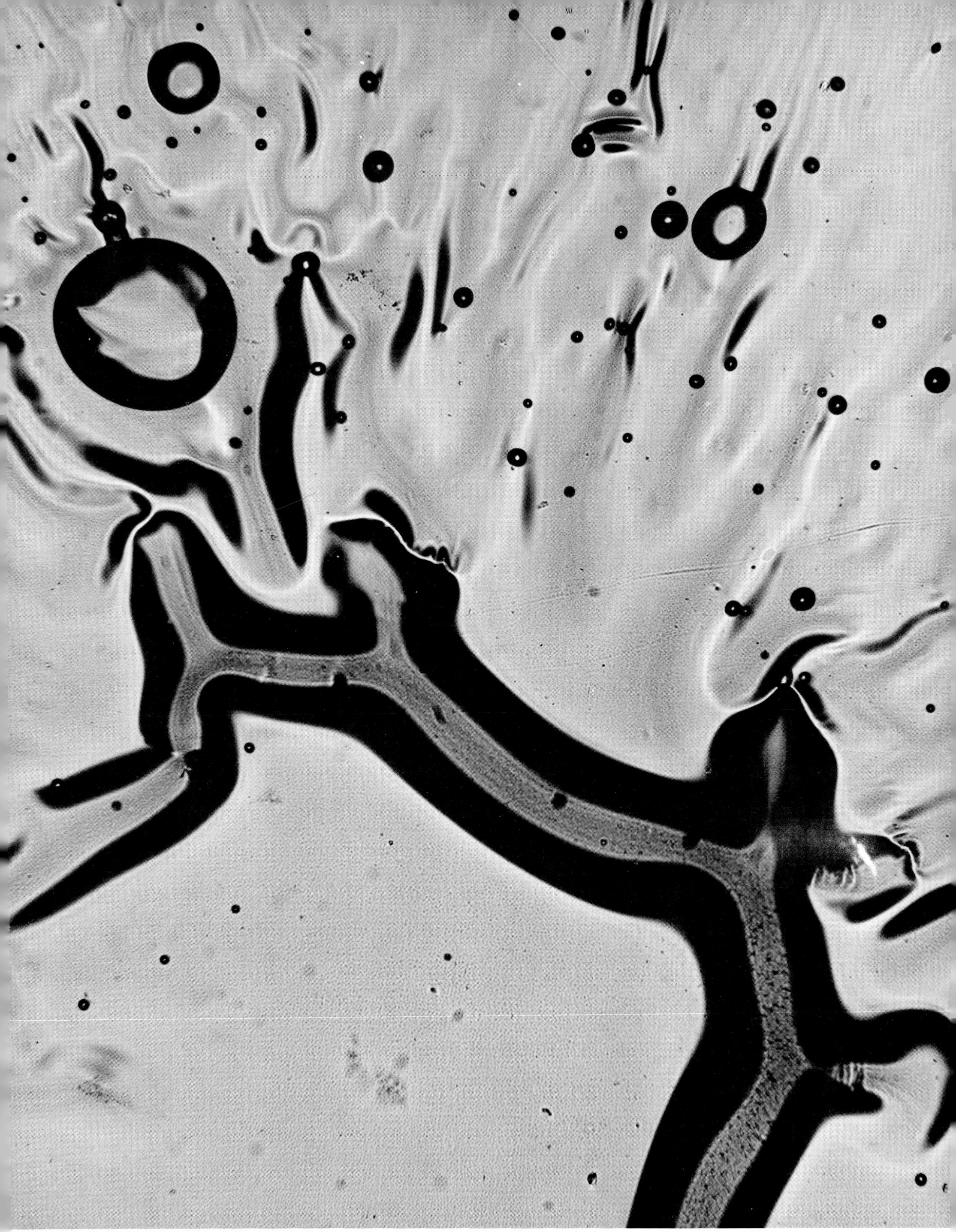

59 Sodium silicate. Magnification 225X.

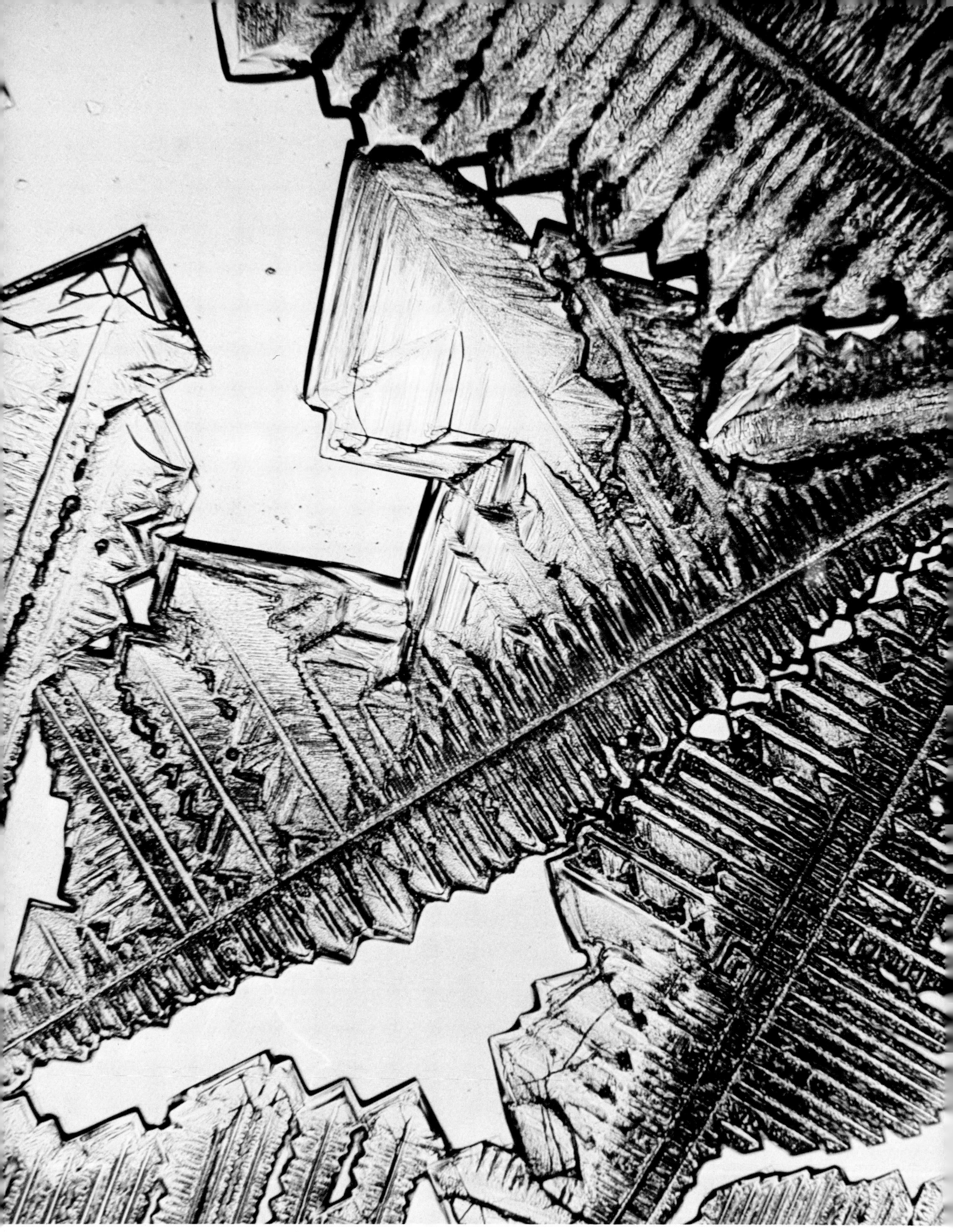

60 Potassium aluminum tartrate. Magnification 225X.

61 DL methionine. Magnification 560X.

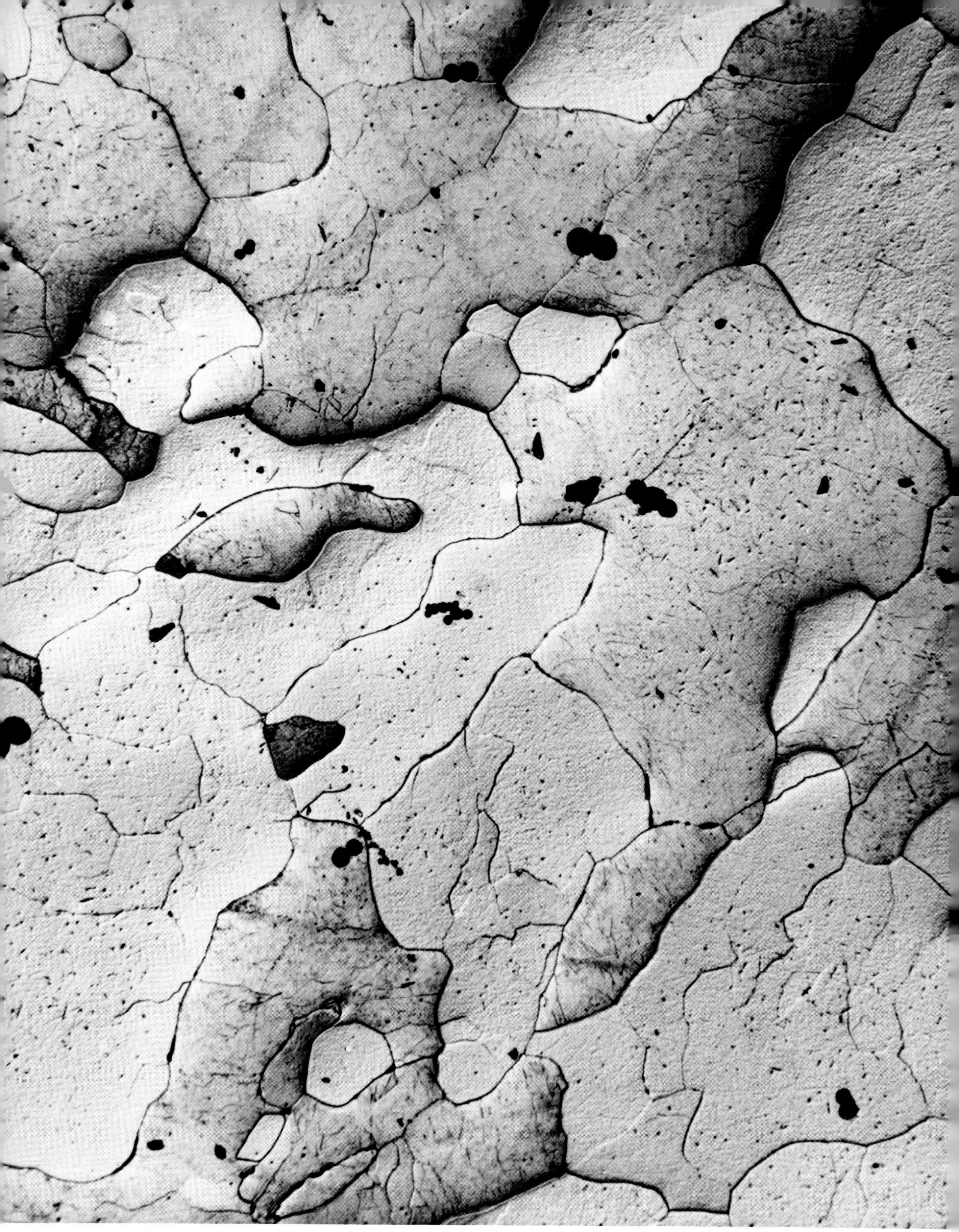

62 Cast iron. Magnification 210X.

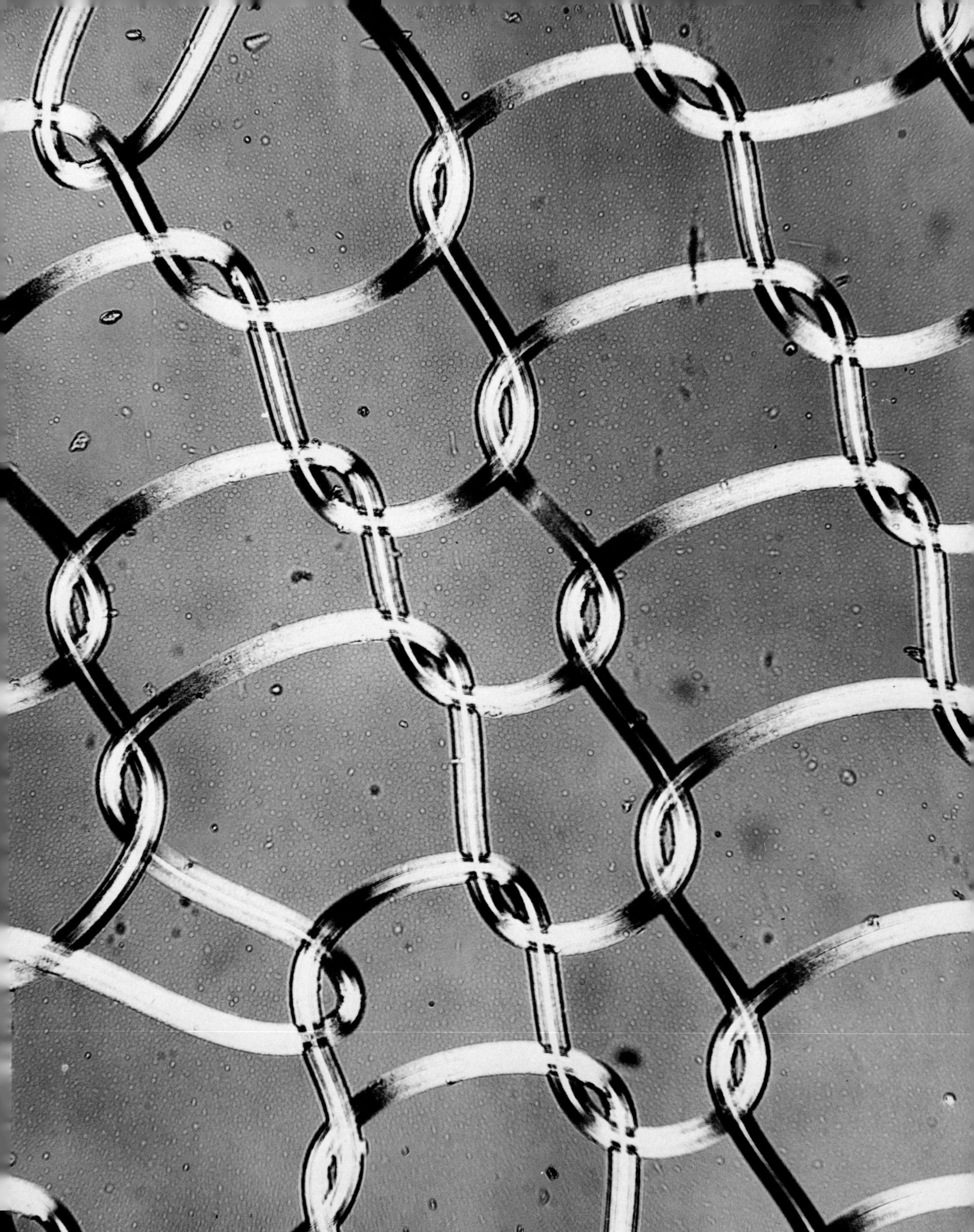

63 Nylon weave. Magnification 875X.

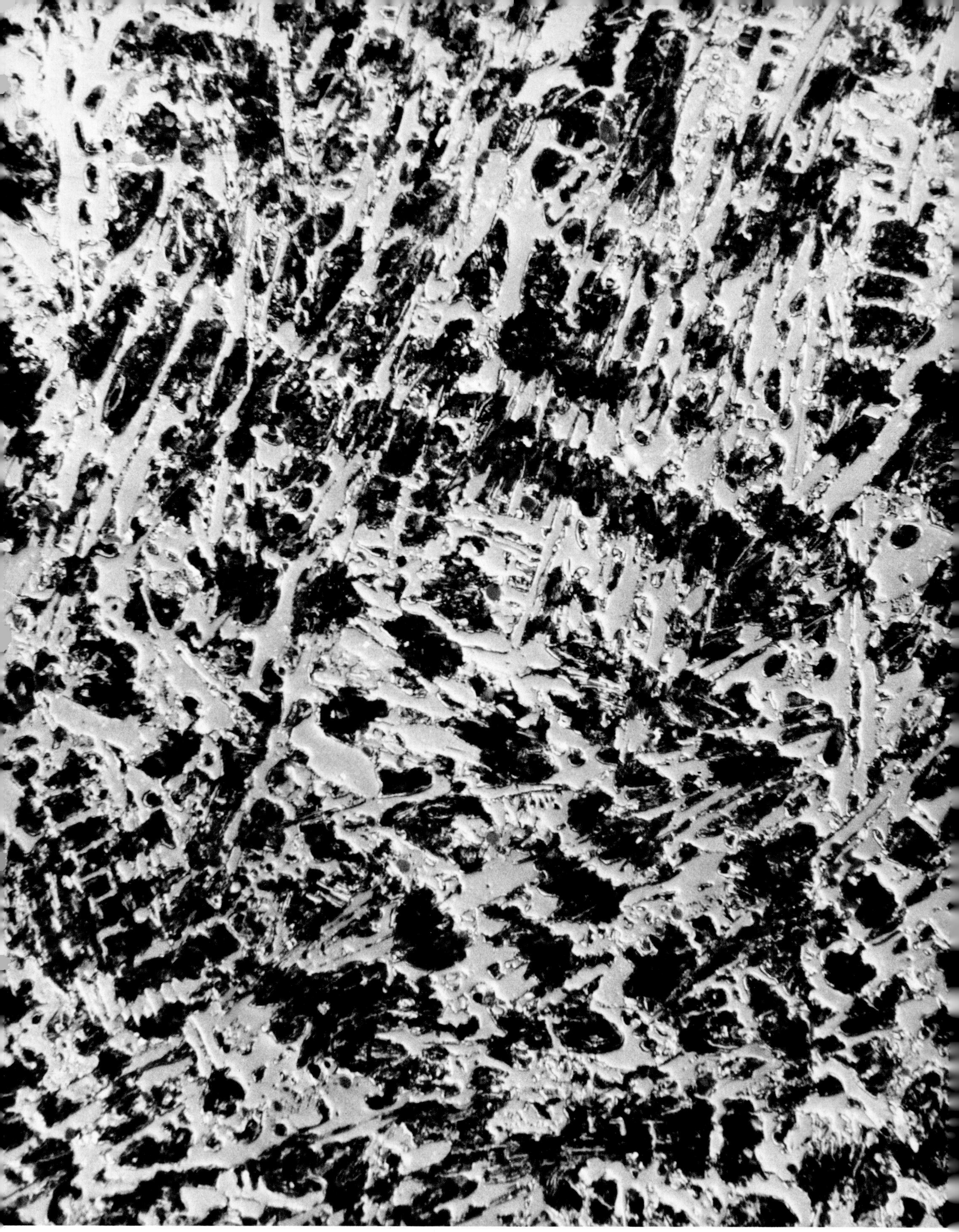

64 Carbon iron alloy. Magnification 210X.

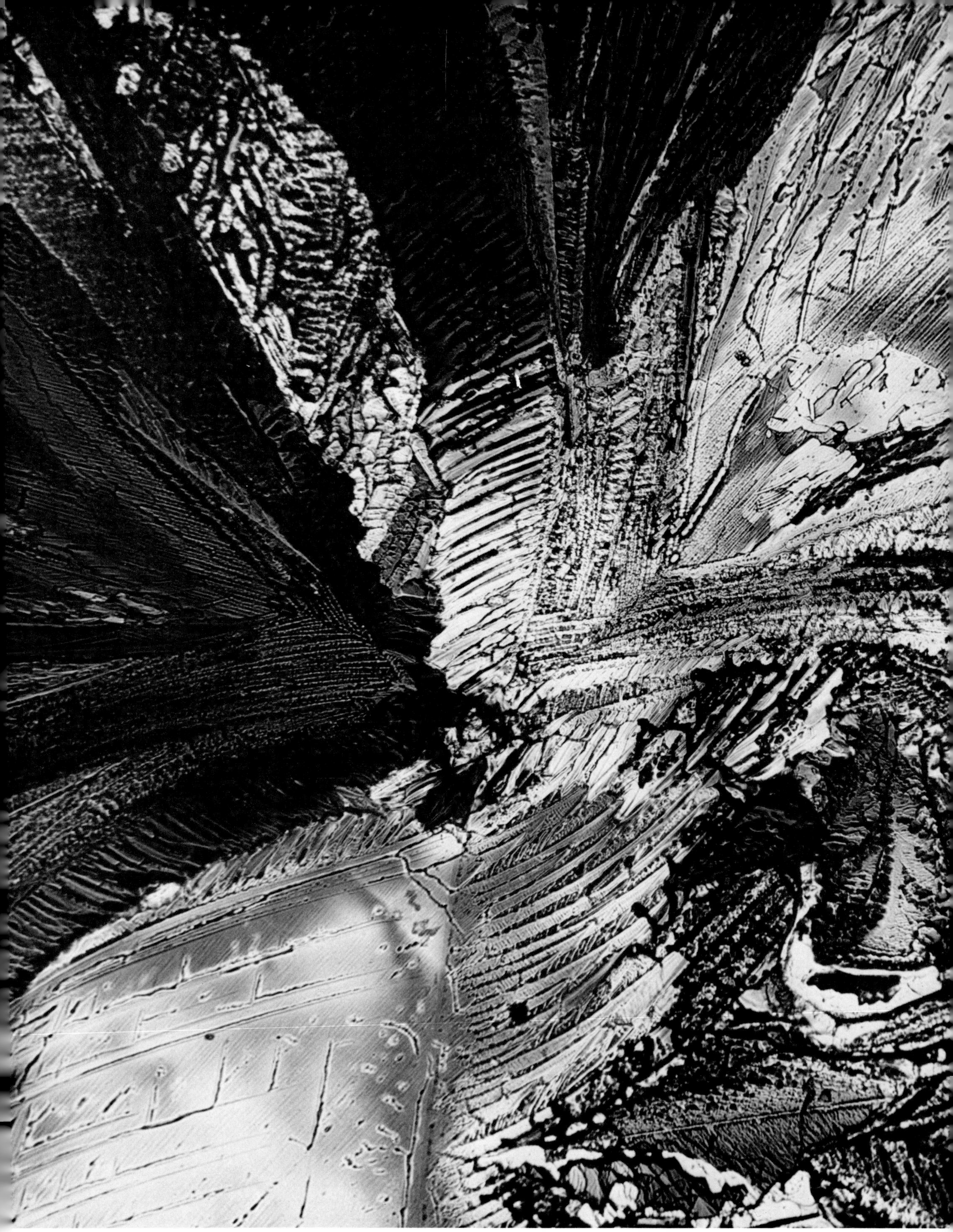

65 Beta-alanine. Magnification 225X.

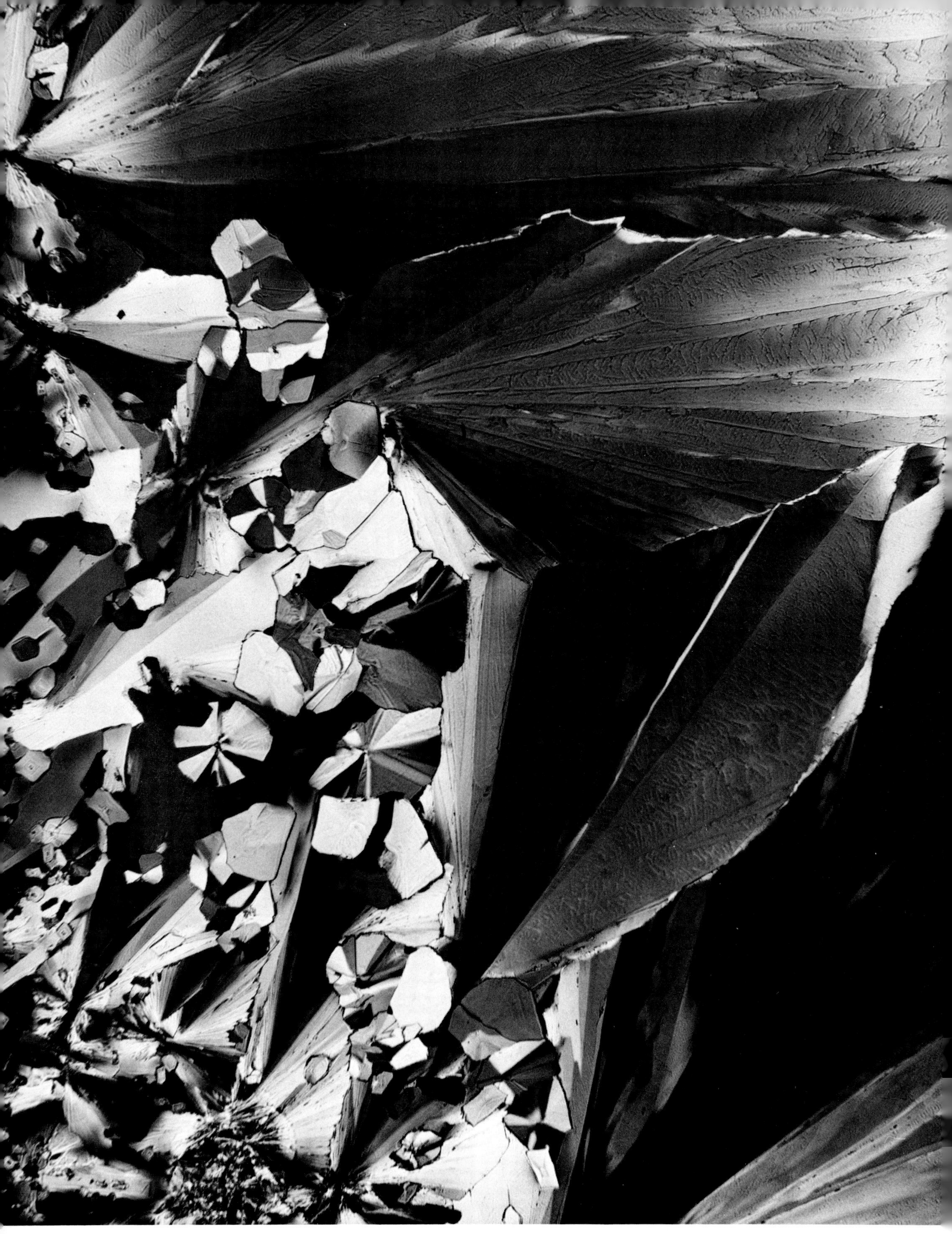

66 Sodium citrate. Magnification 210X.

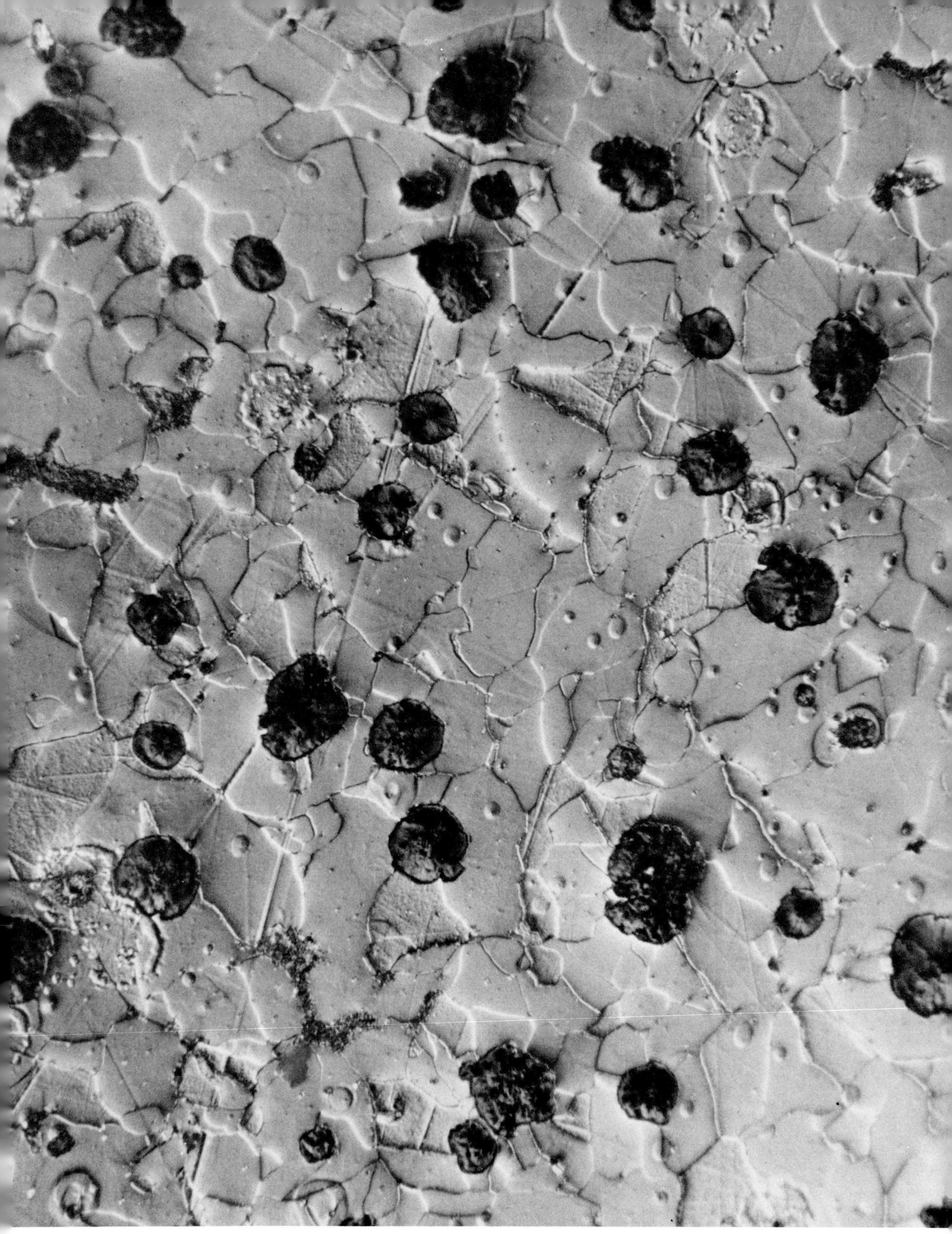

67 Nodular cast iron. Magnification 700X.

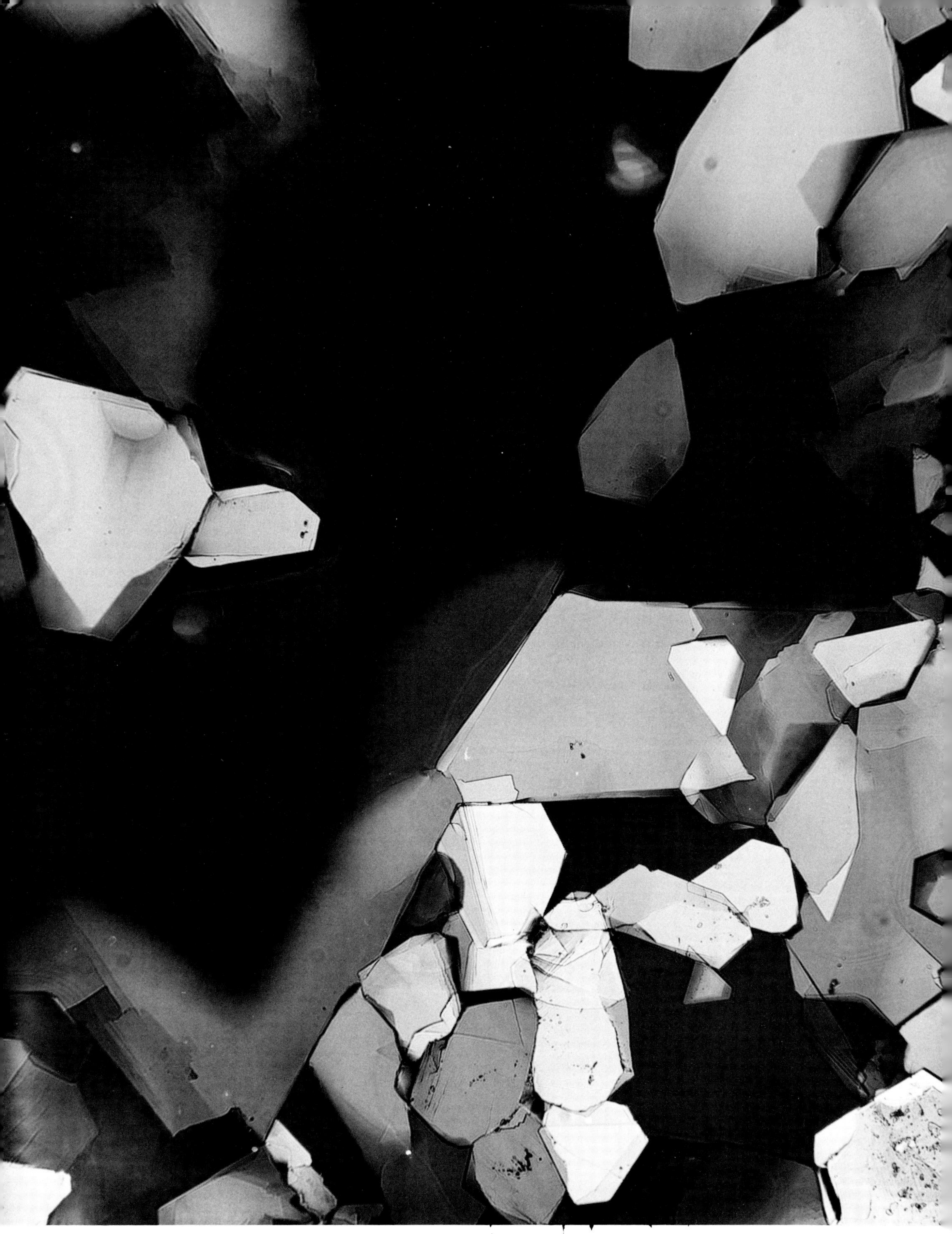

68 DL methionine. Magnification 560X.

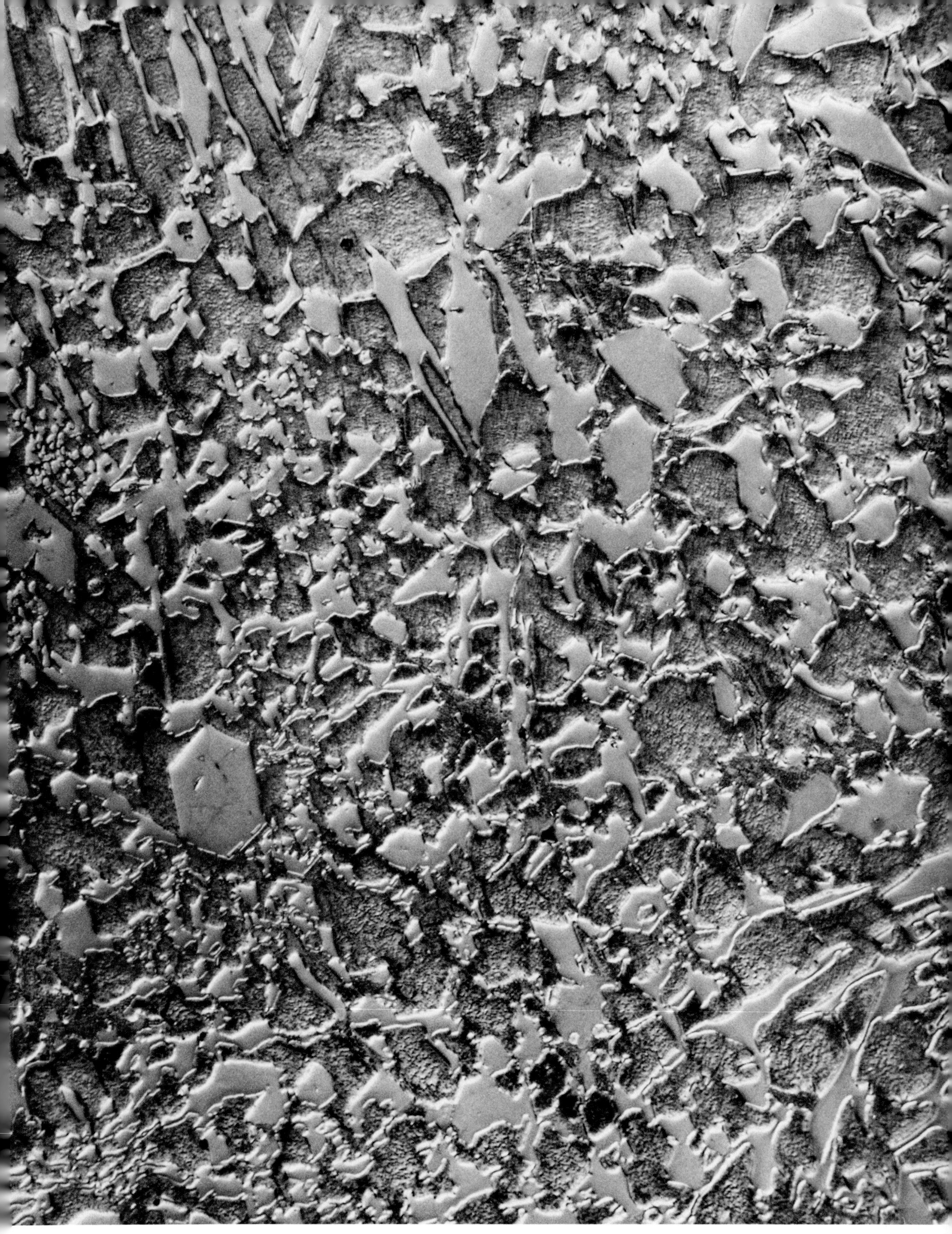

69 White iron containing 27% Cr and 5% Mo. Magnification 875X.

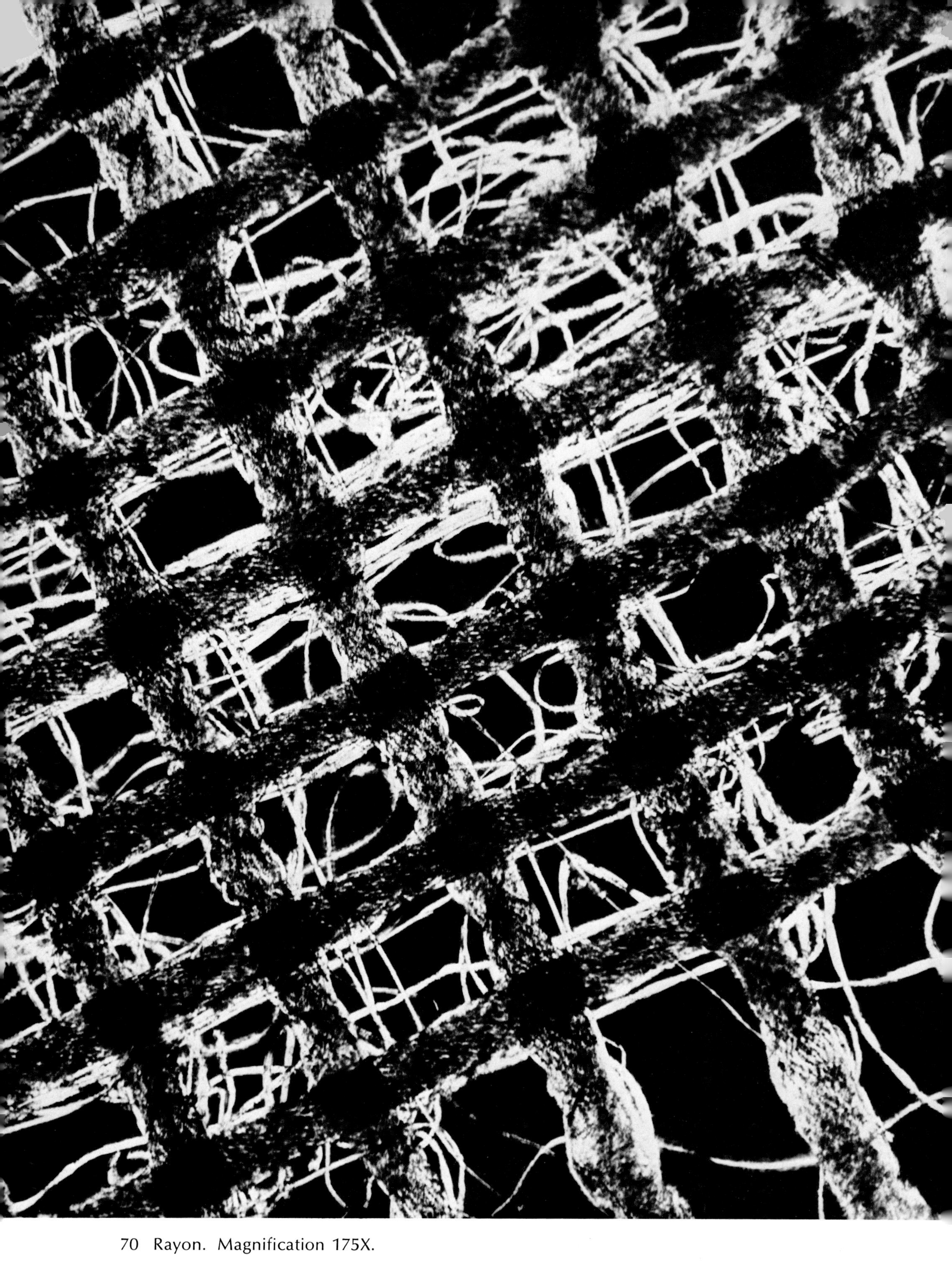

70 Rayon. Magnification 175X.

71 Plastic material. Magnification 560X.

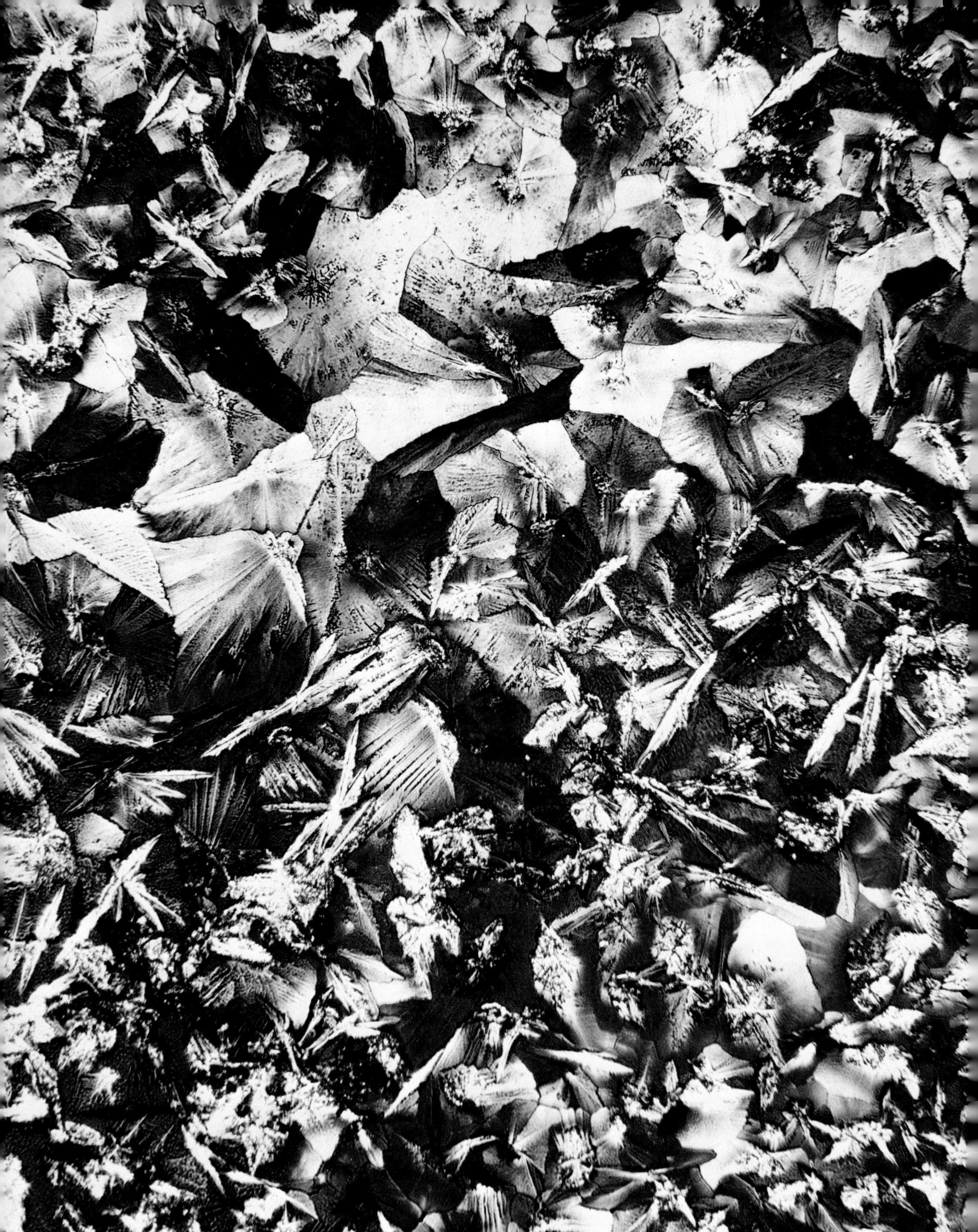

72 Copper sulfate. Magnification 210X.

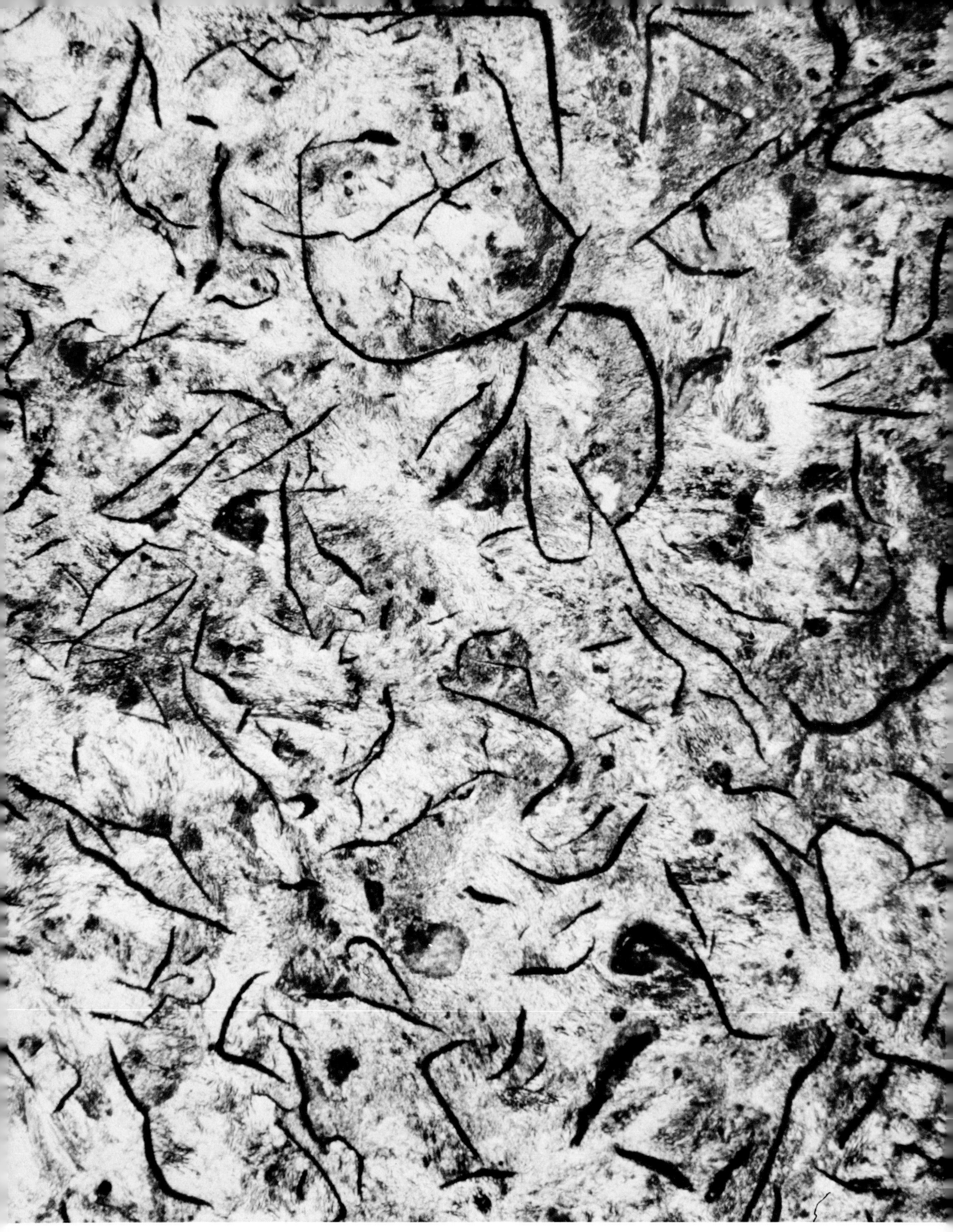

73 Ductile iron. Magnification 875X.

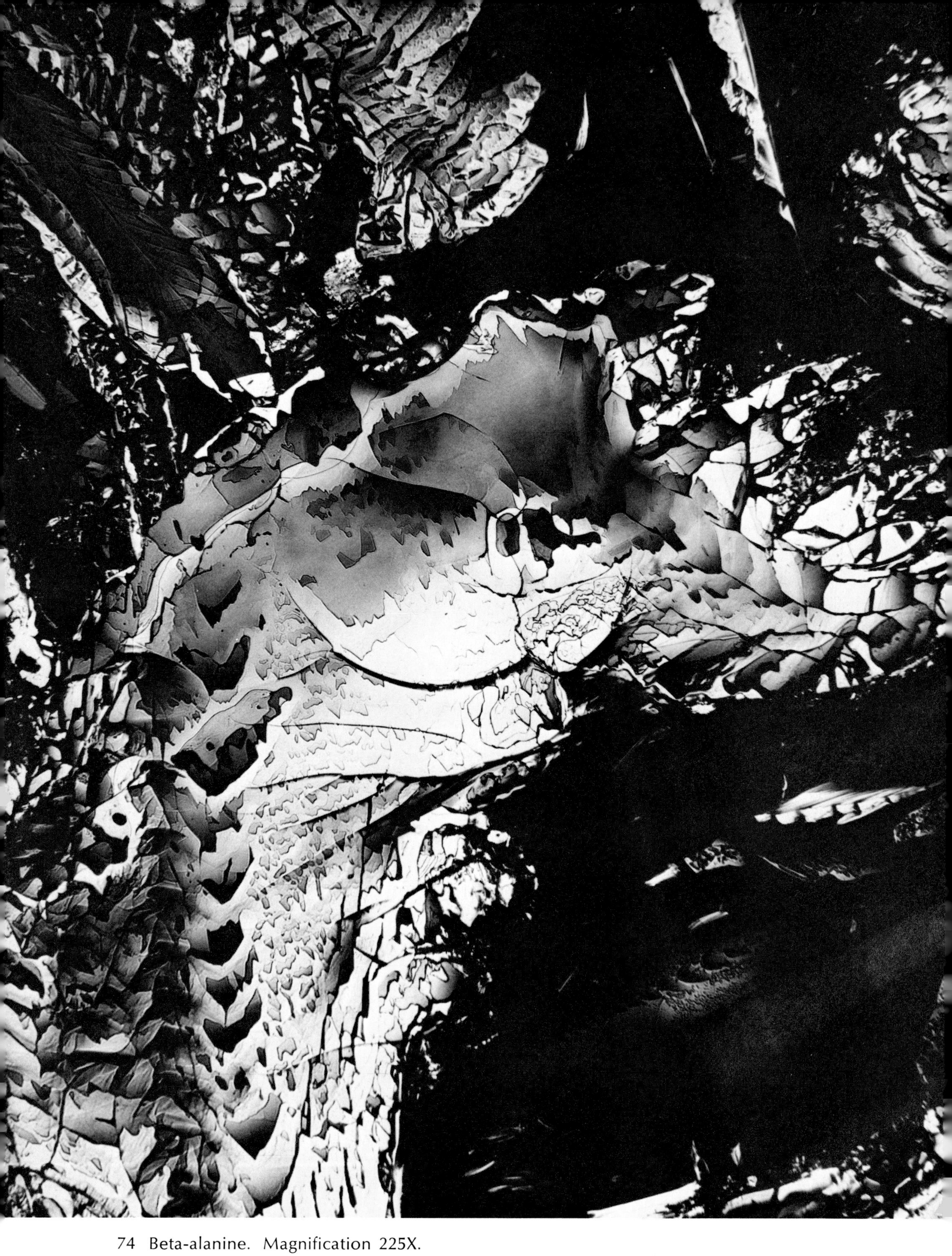

74 Beta-alanine. Magnification 225X.

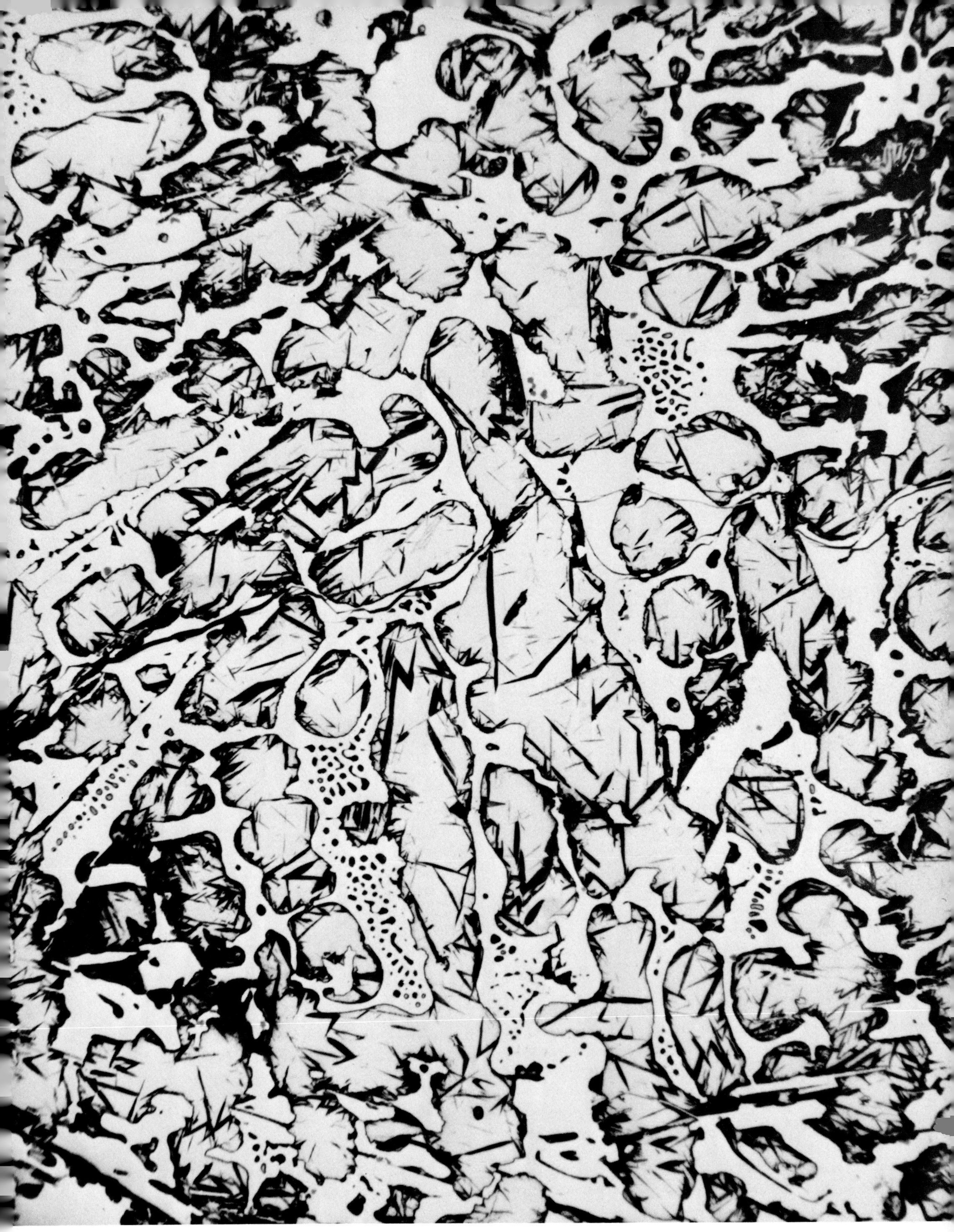

75 Cast steel. Magnification 700X.

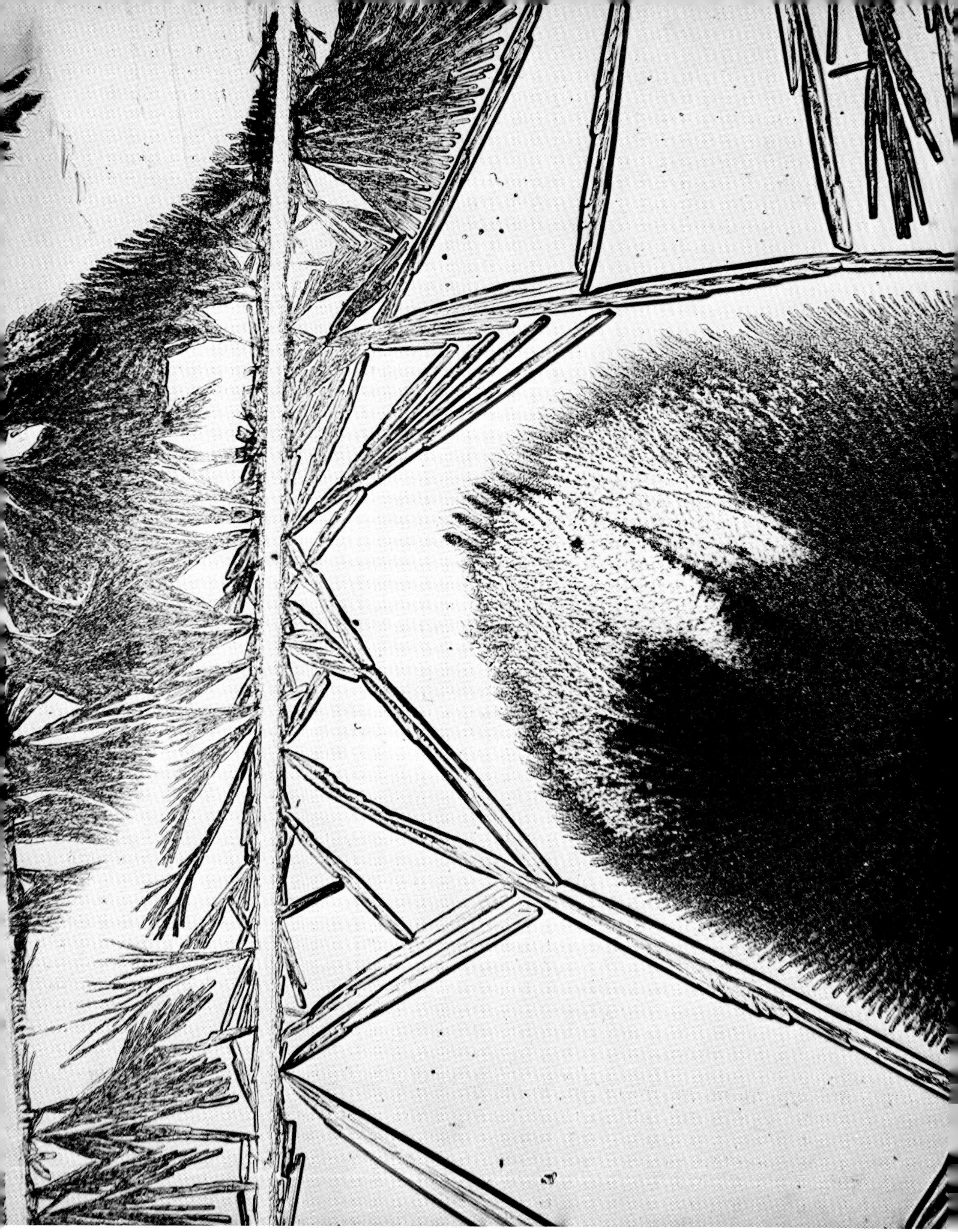

76 Camphor and cyclohexone. Magnification 225X.

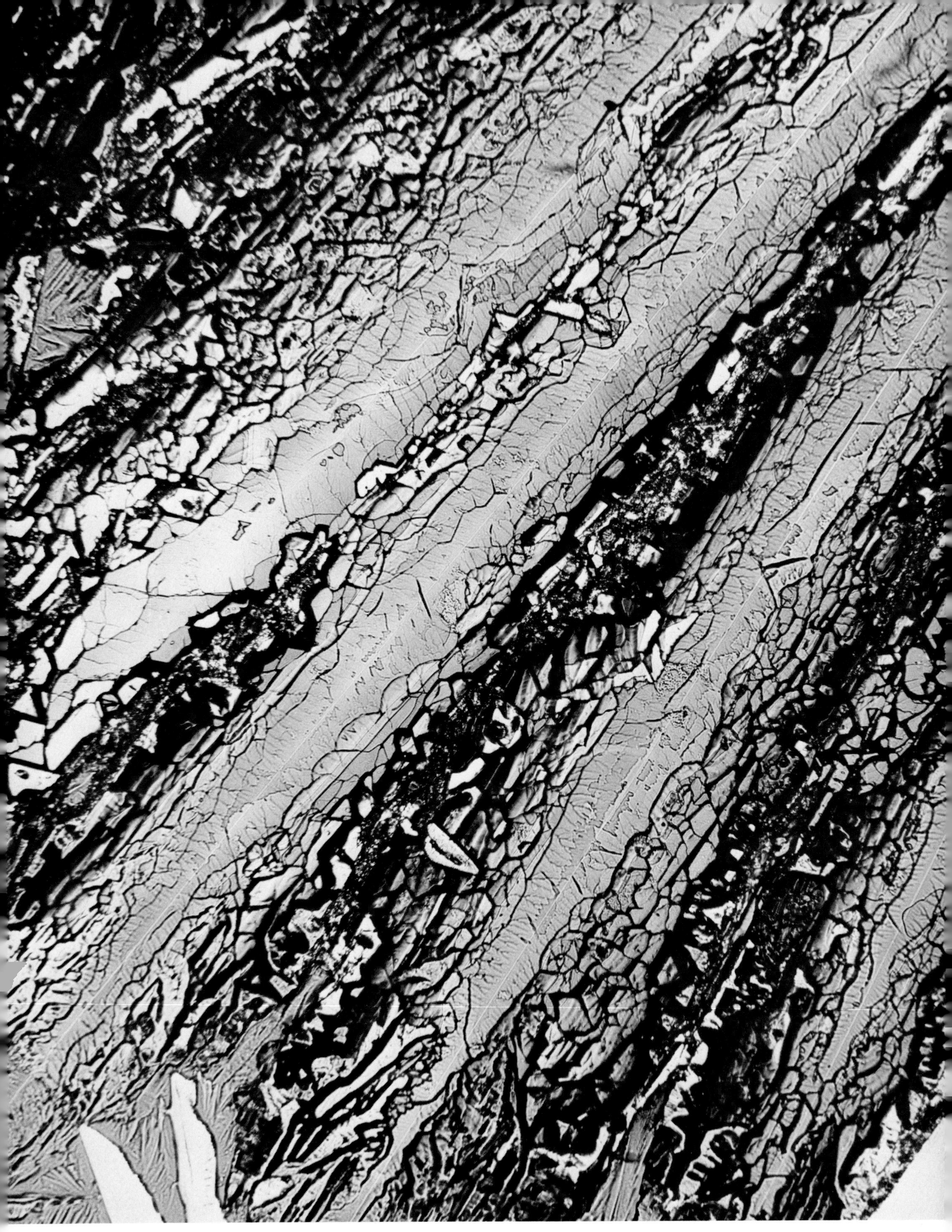

77 Beta-alanine. Magnification 225X.

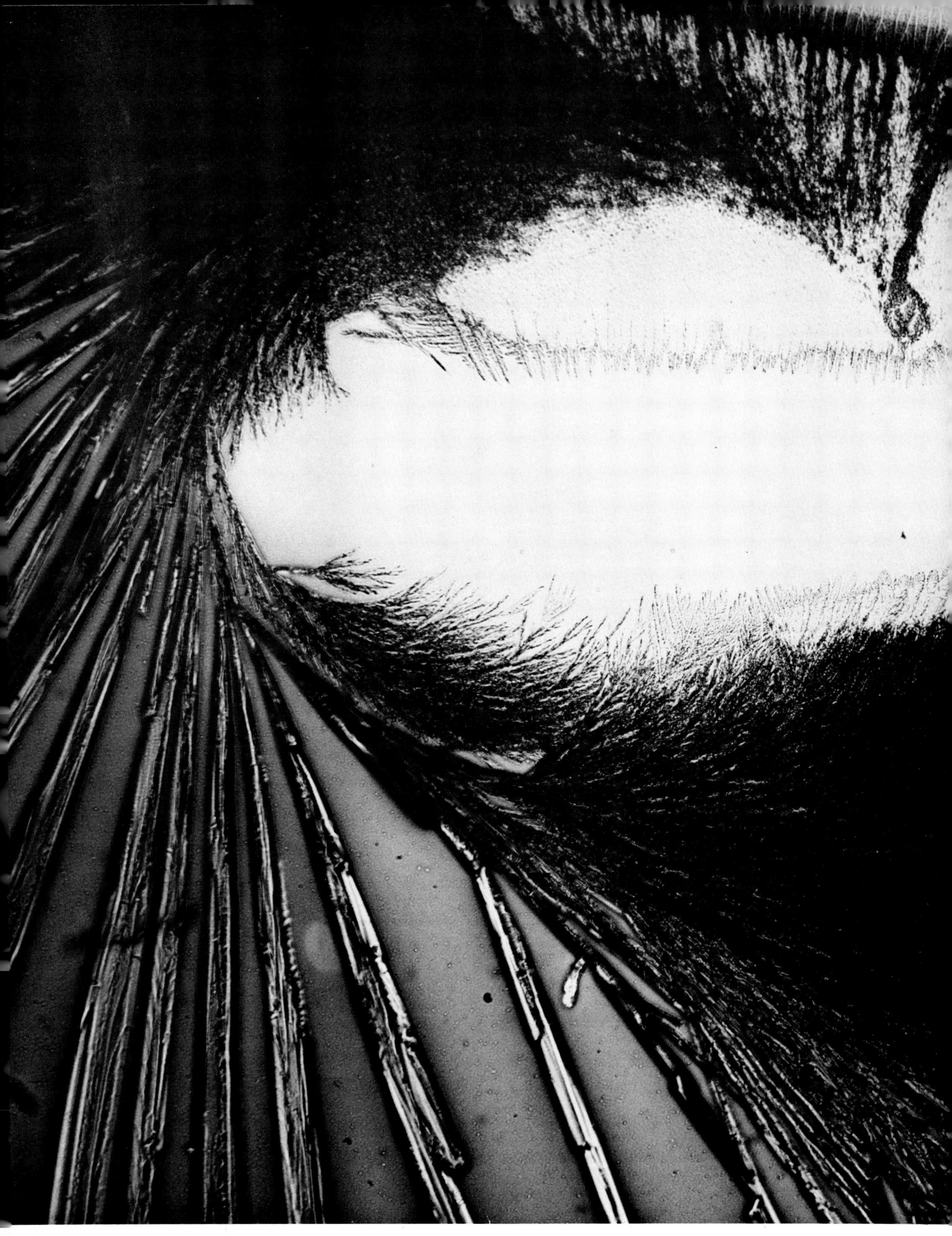

78 Camphor and cyclohexanone. Magnification 225X.

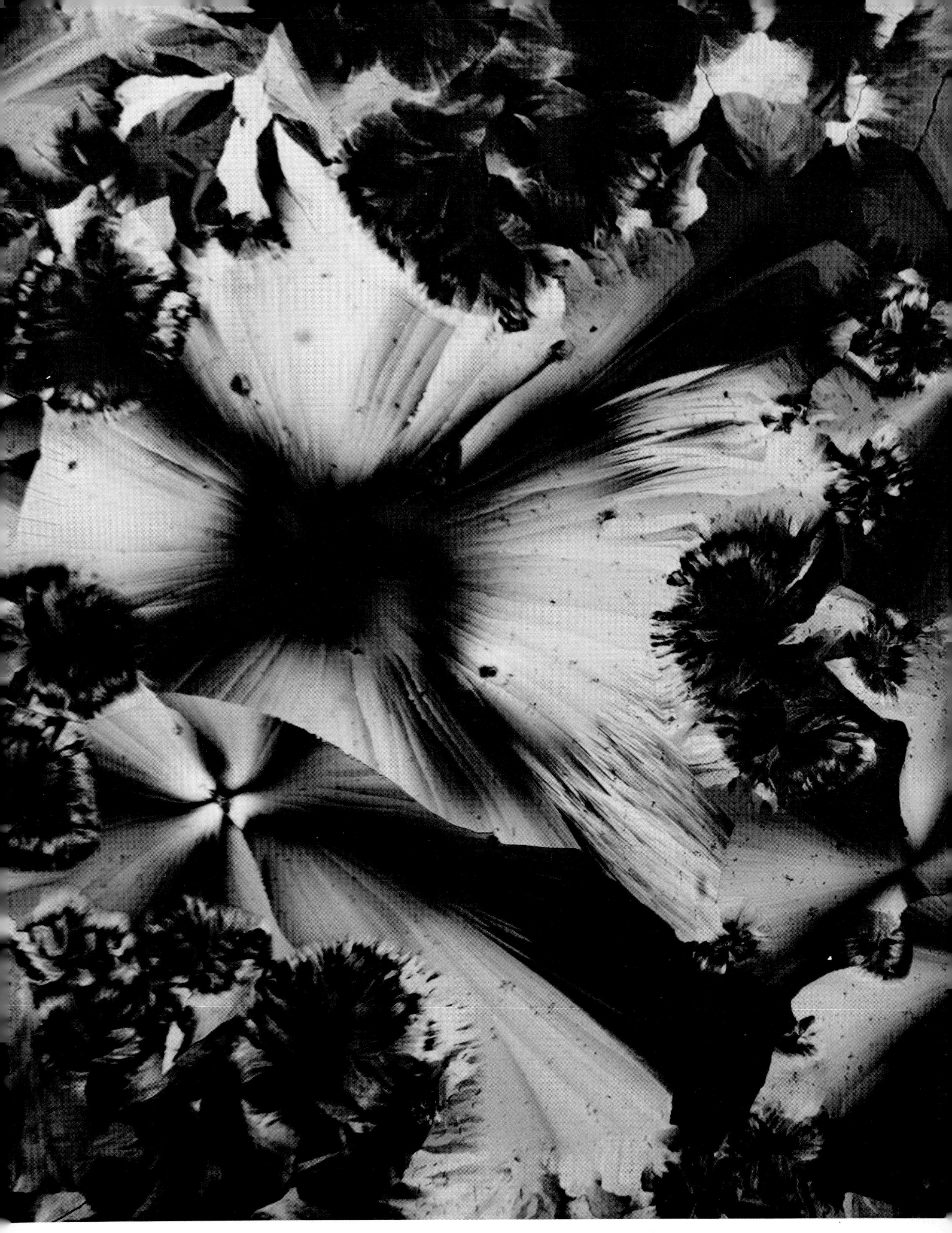

79 Magnesium sulfate (Epsom salt). Magnification 225X.

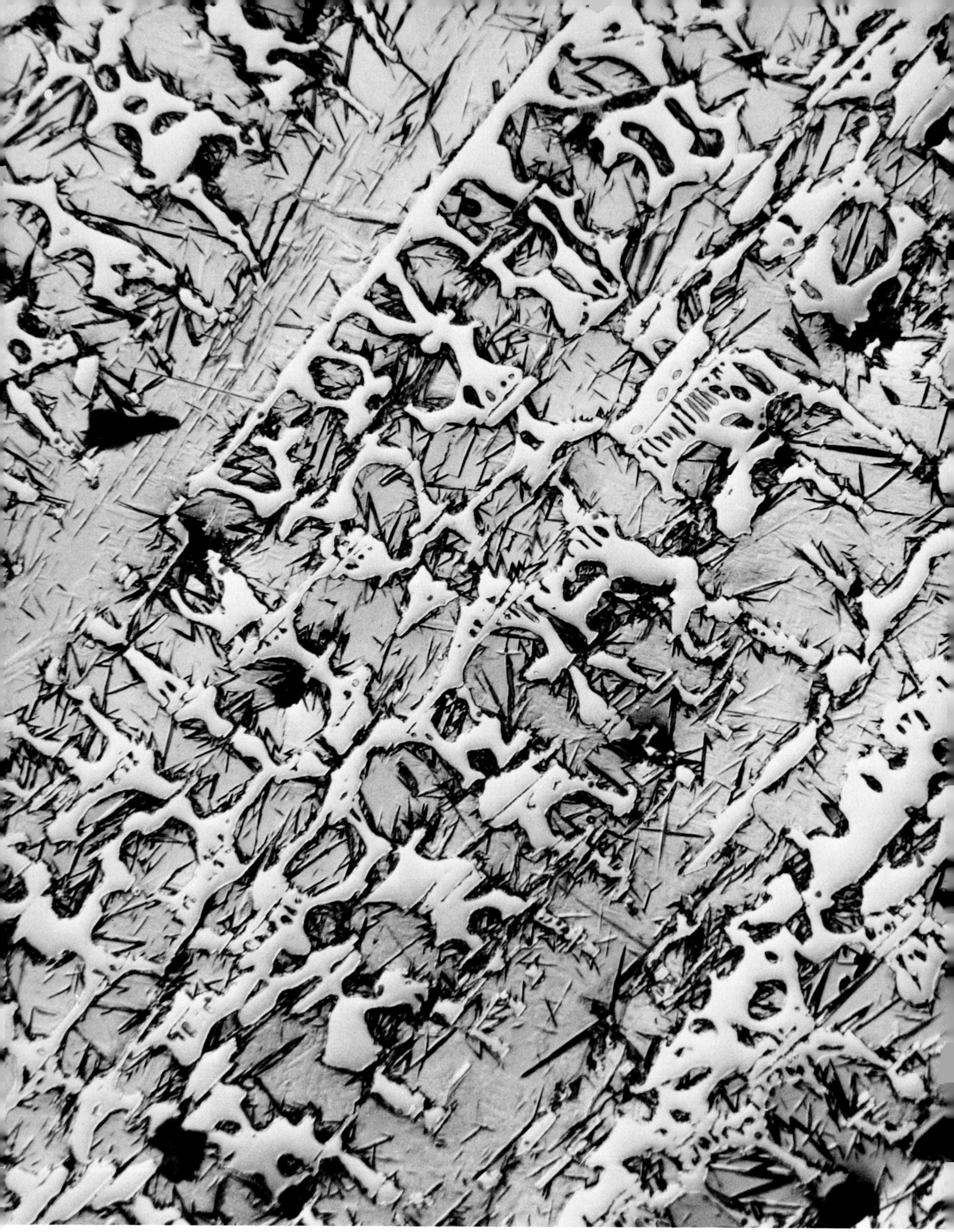

80 Cast steel with silicon. Magnification 700X.

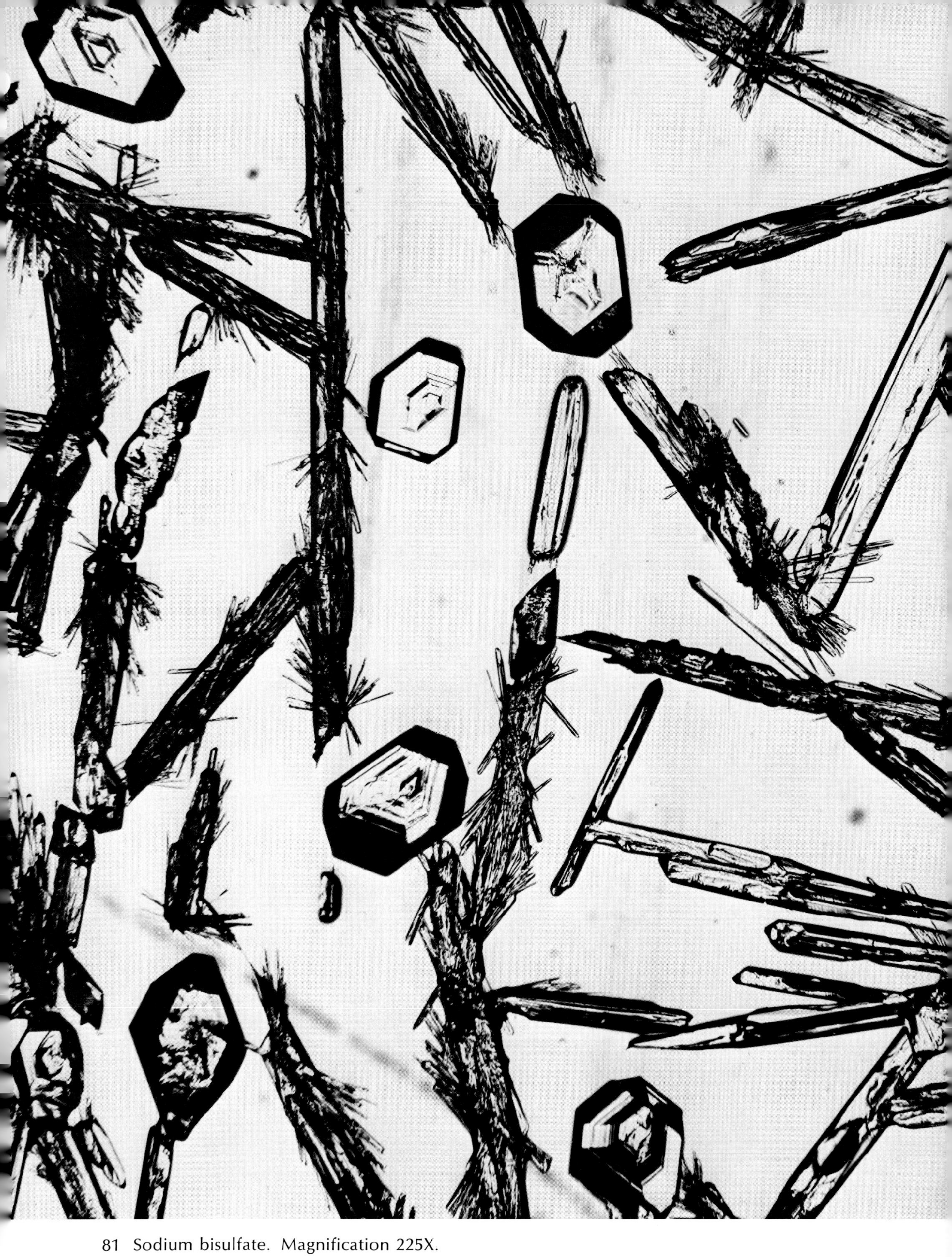

81 Sodium bisulfate. Magnification 225X.

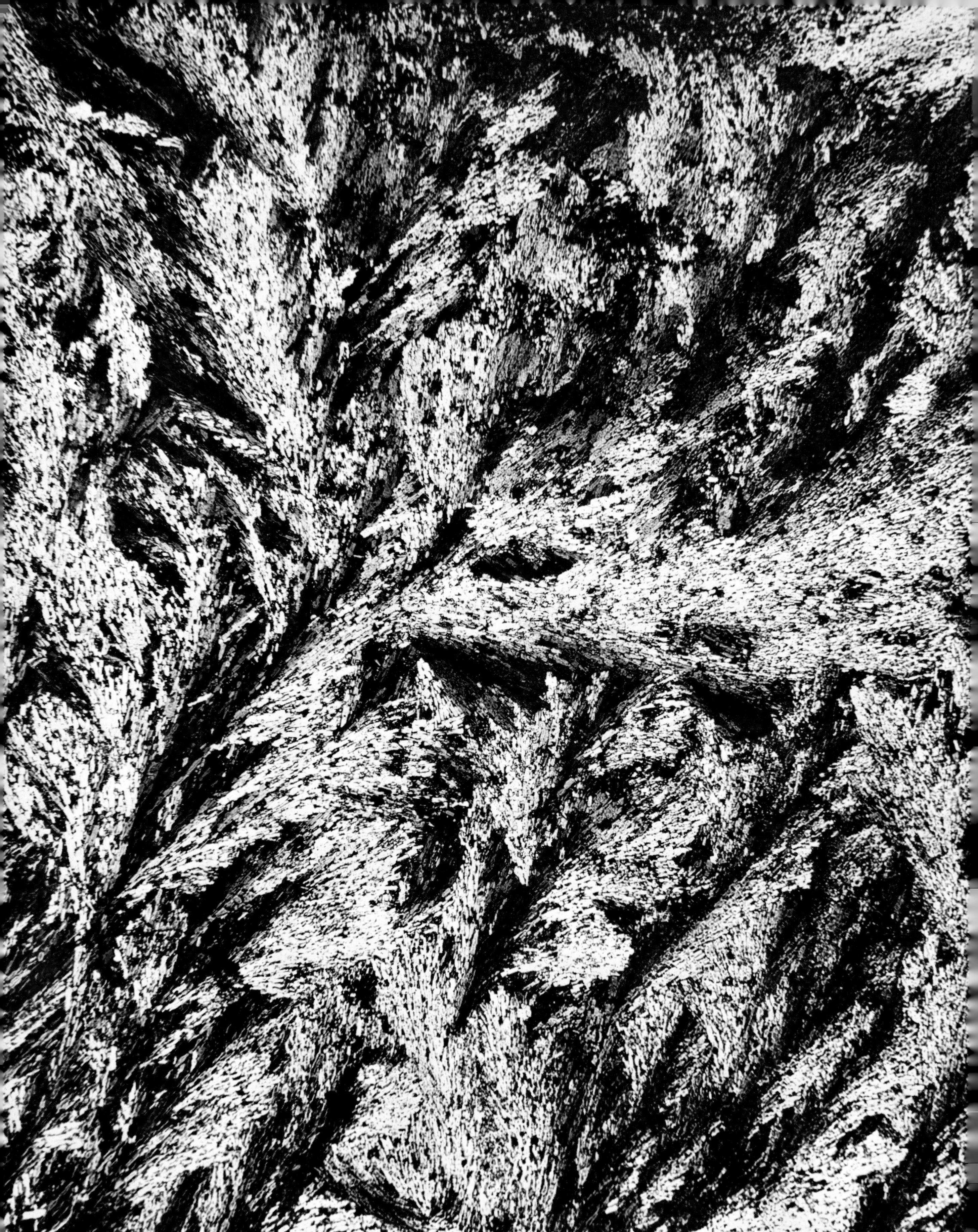

82 Strontium bromide. Magnification 210X.

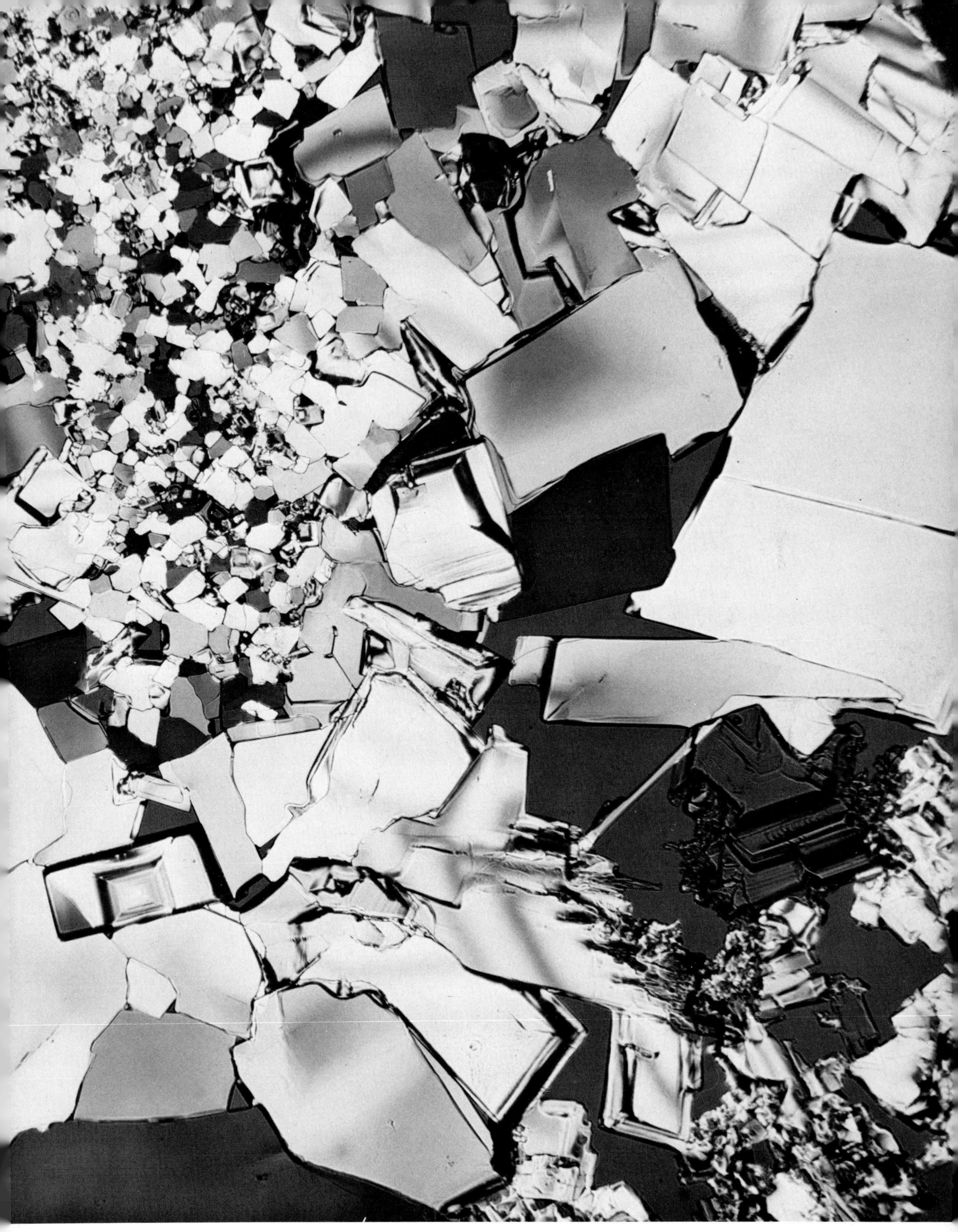

83 Thymol. Magnification 210X.

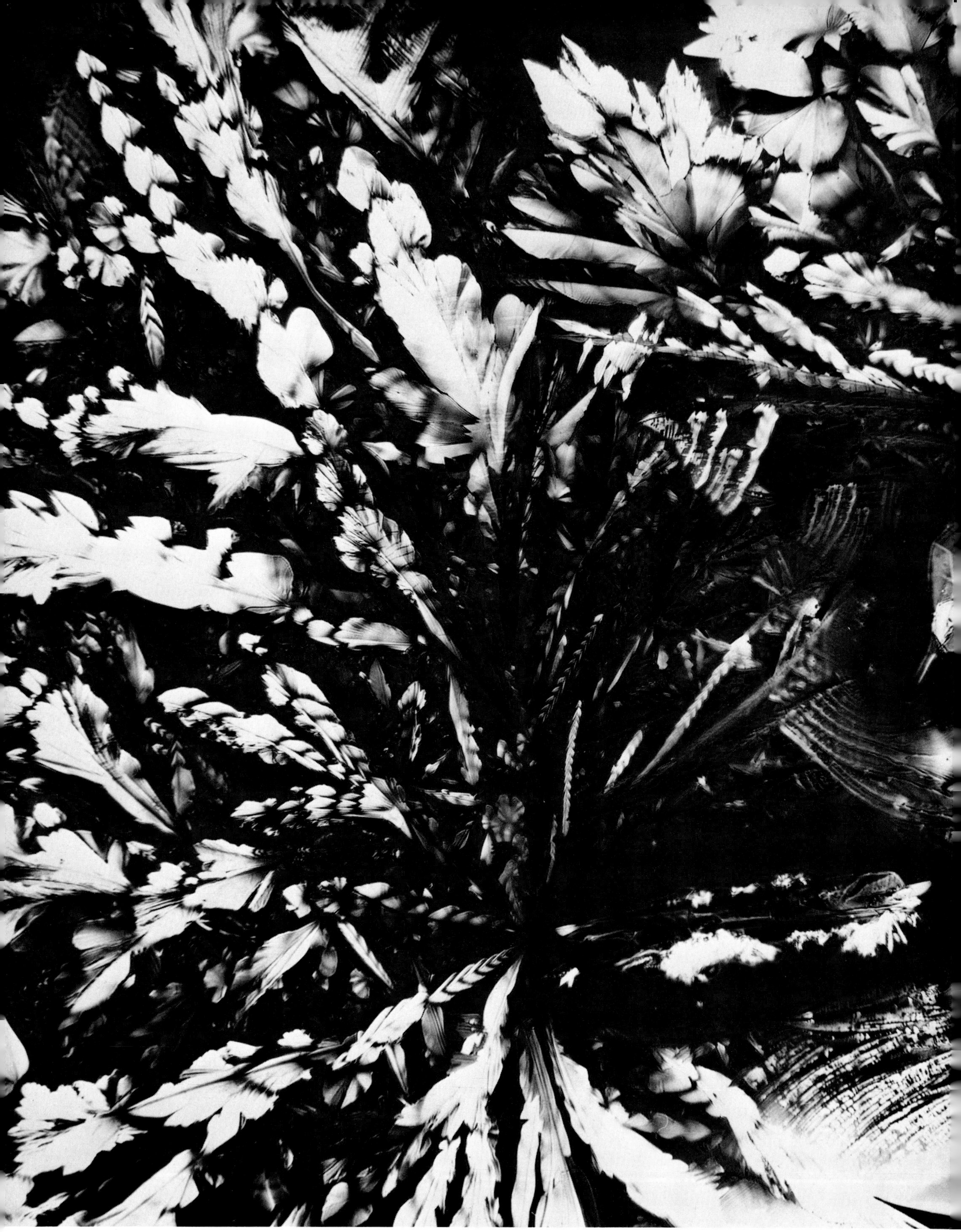

84 DL-serine. Magnification 560X.

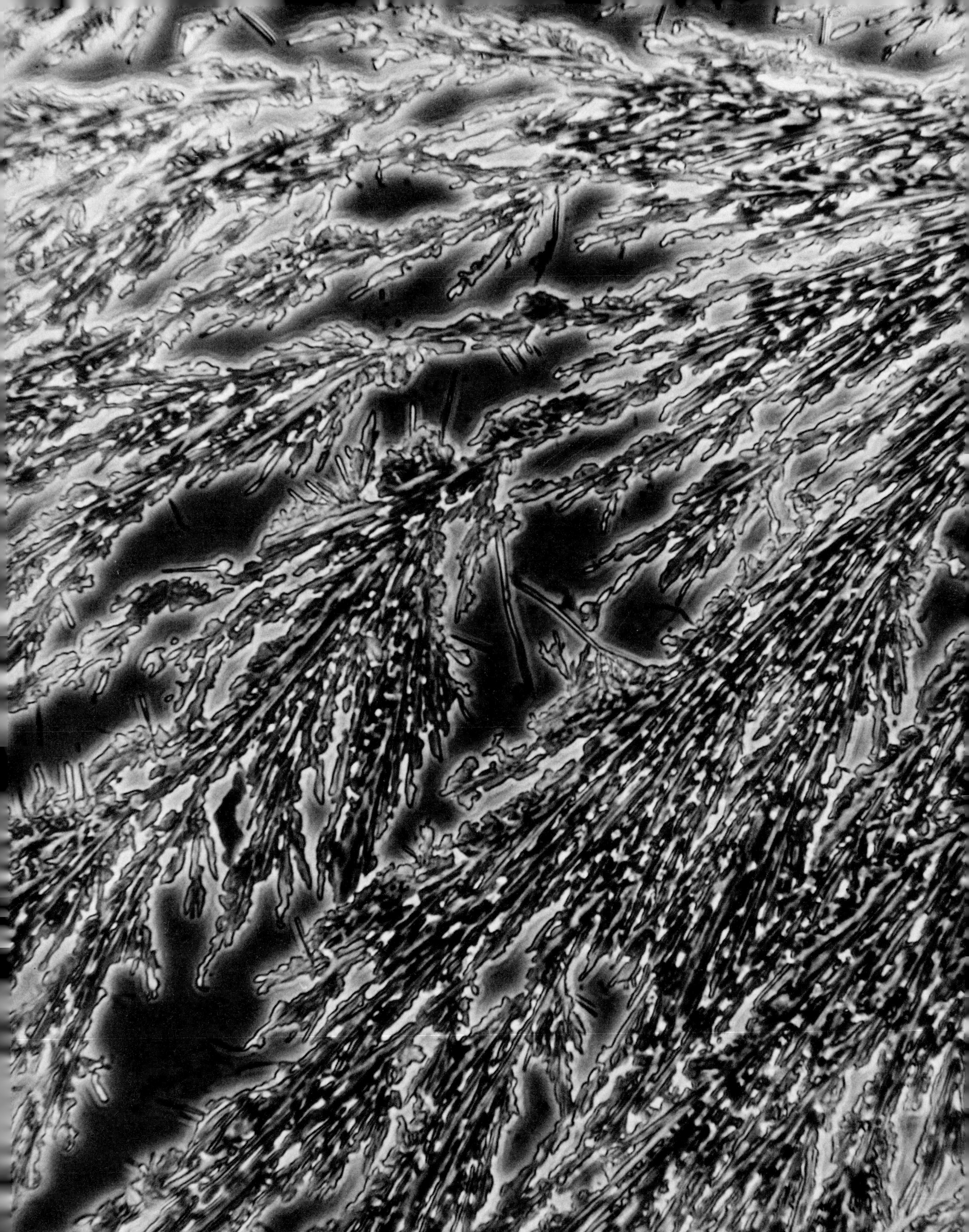

85 Potassium ferricyanide. Magnification 440X.

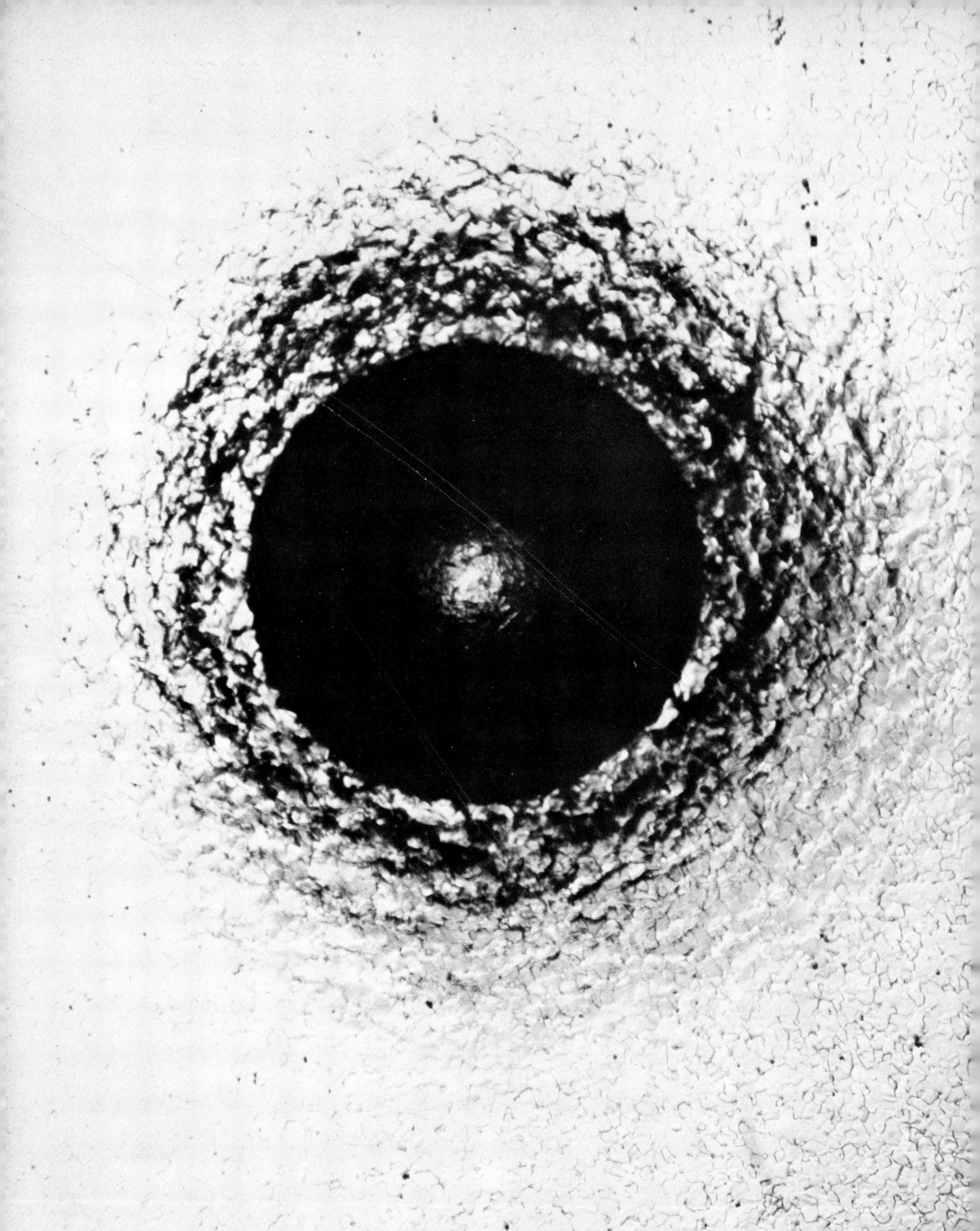

86 Low-carbon annealed steel. Magnification 225X.

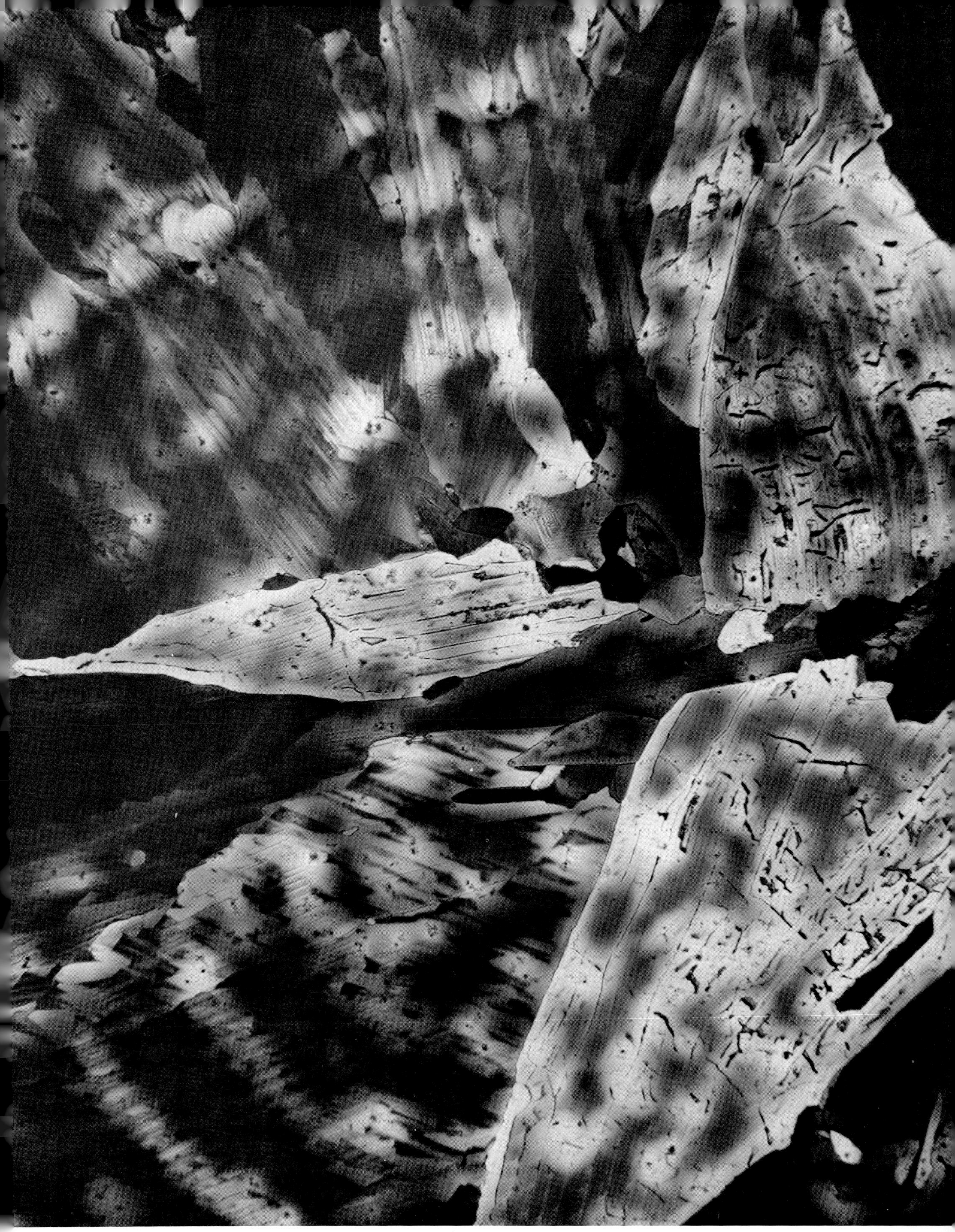

87 Nickel sulfate. Magnification 225X.

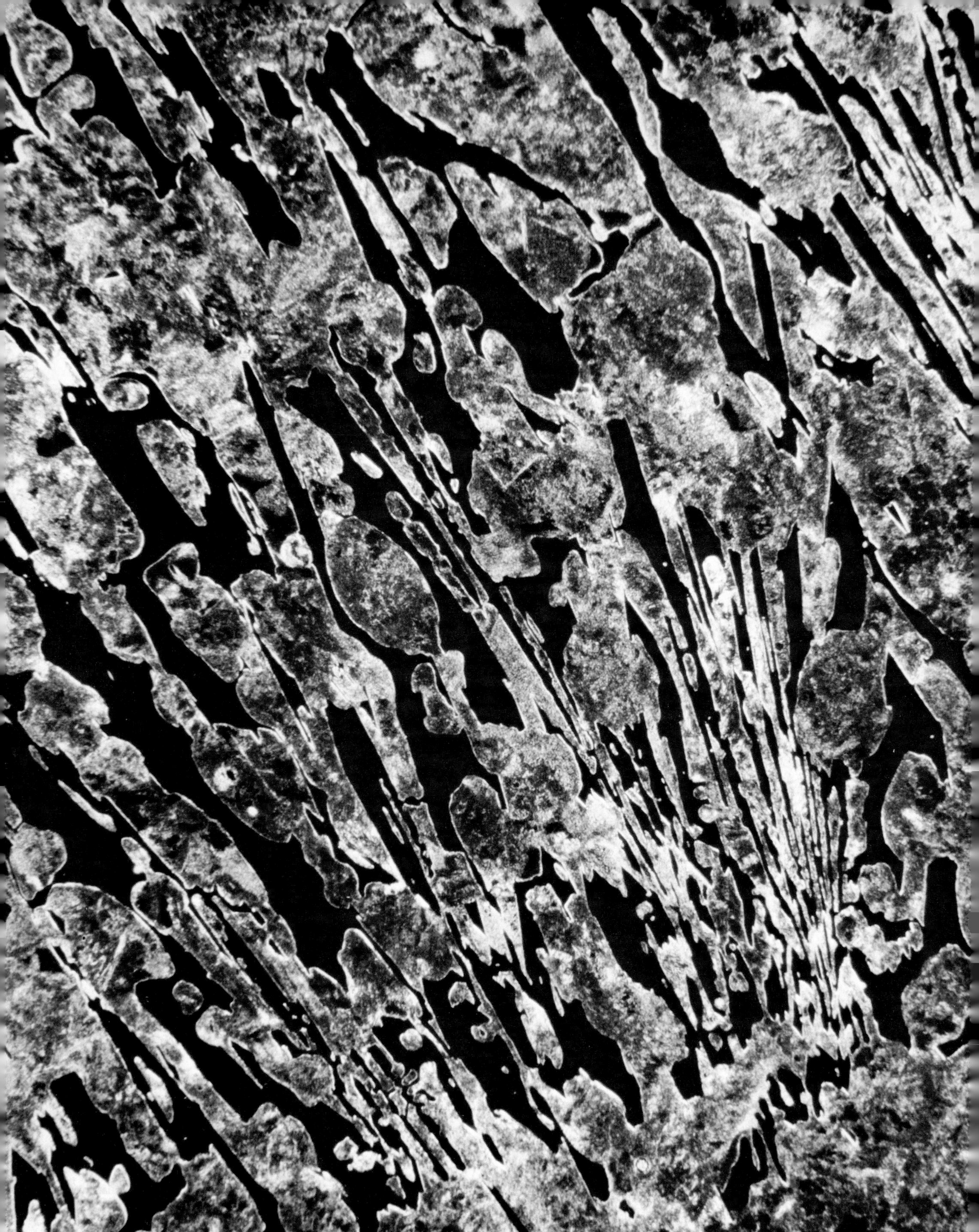

88 White iron containing 13% Cr, 0.7% Mo, 1.2% Cu. Magnification 875X.

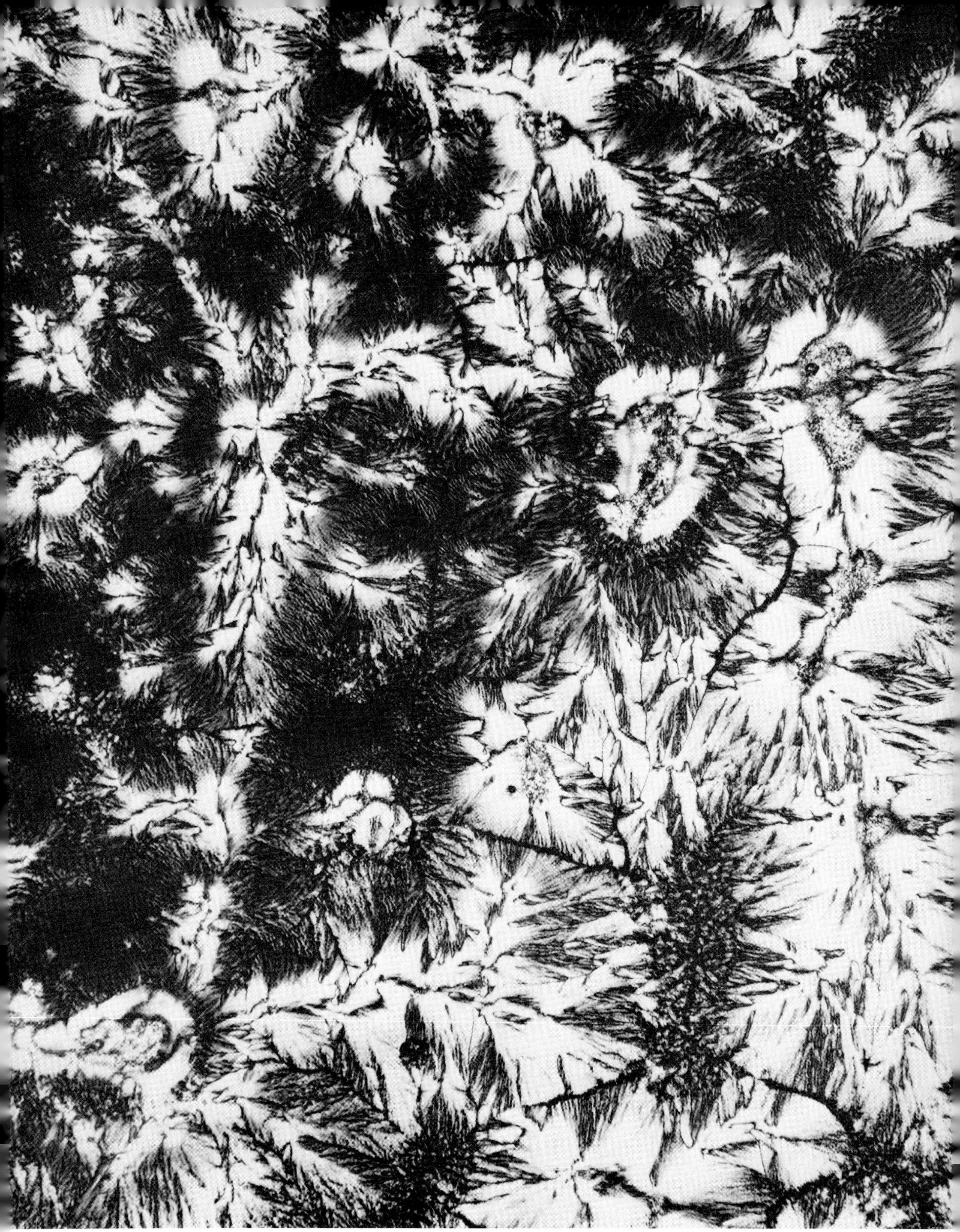

89 Vitamin C. Magnification 560X.

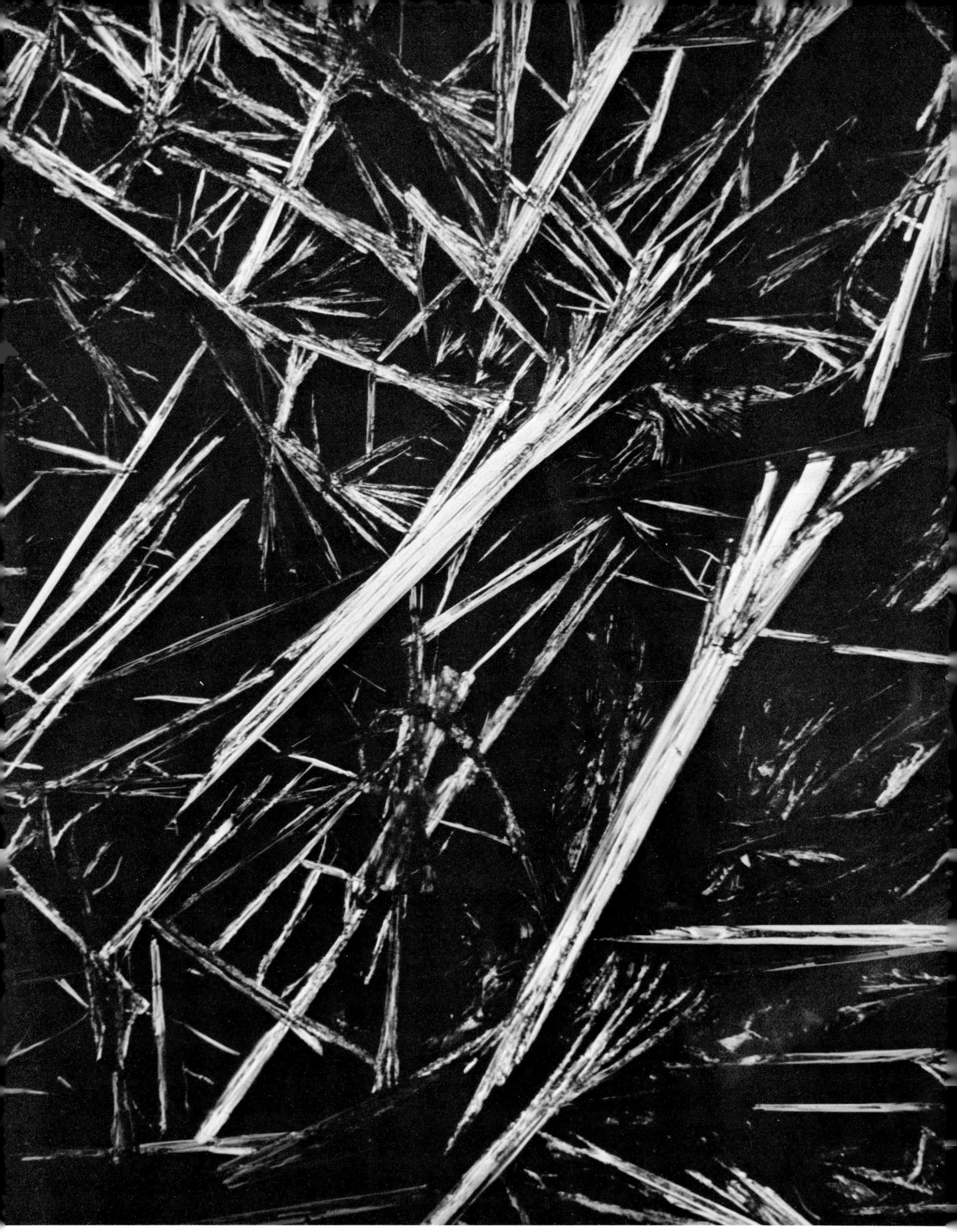

90 Salicylic acid. Magnification 225X.

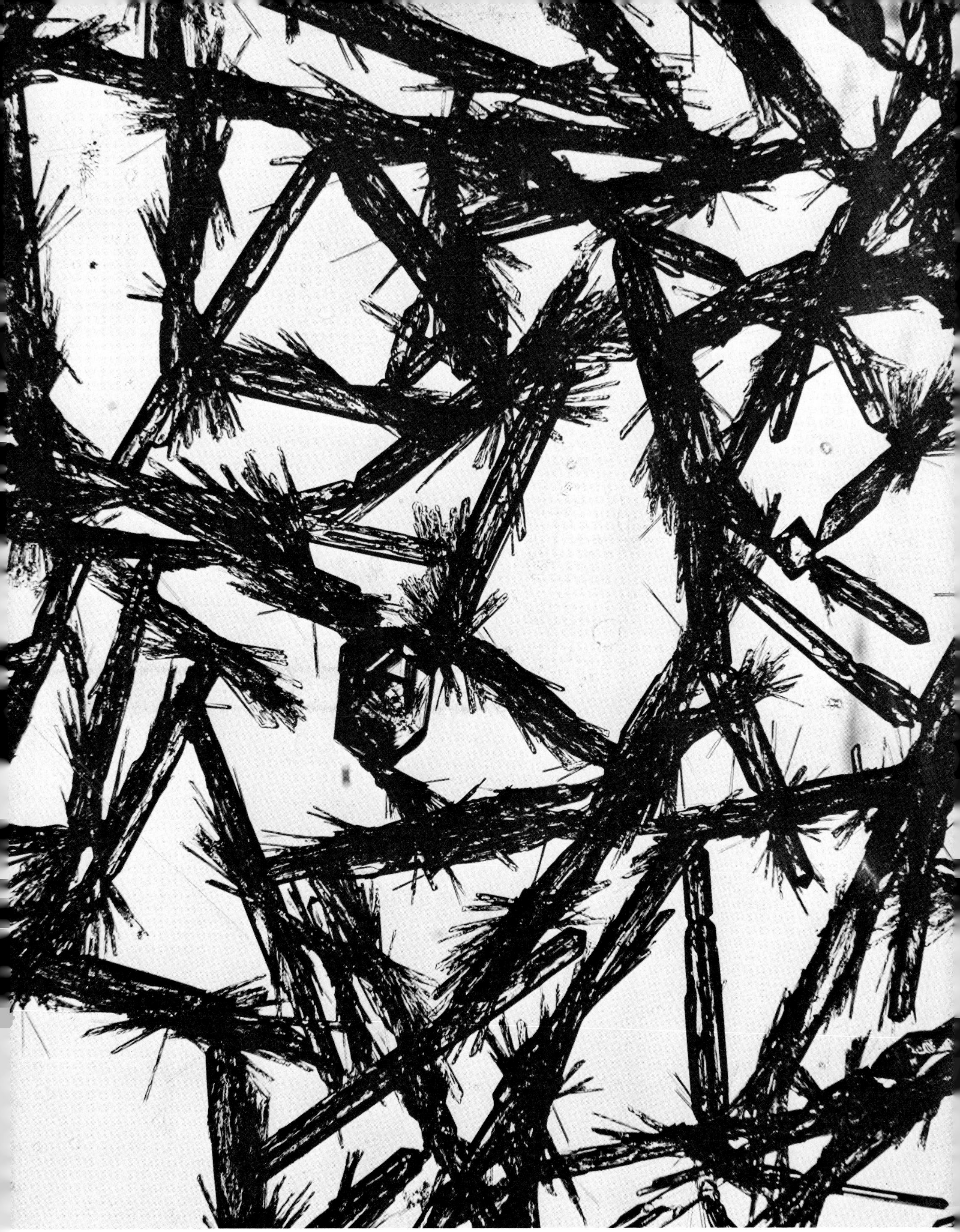

91 Sodium bisulfate. Magnification 420X.

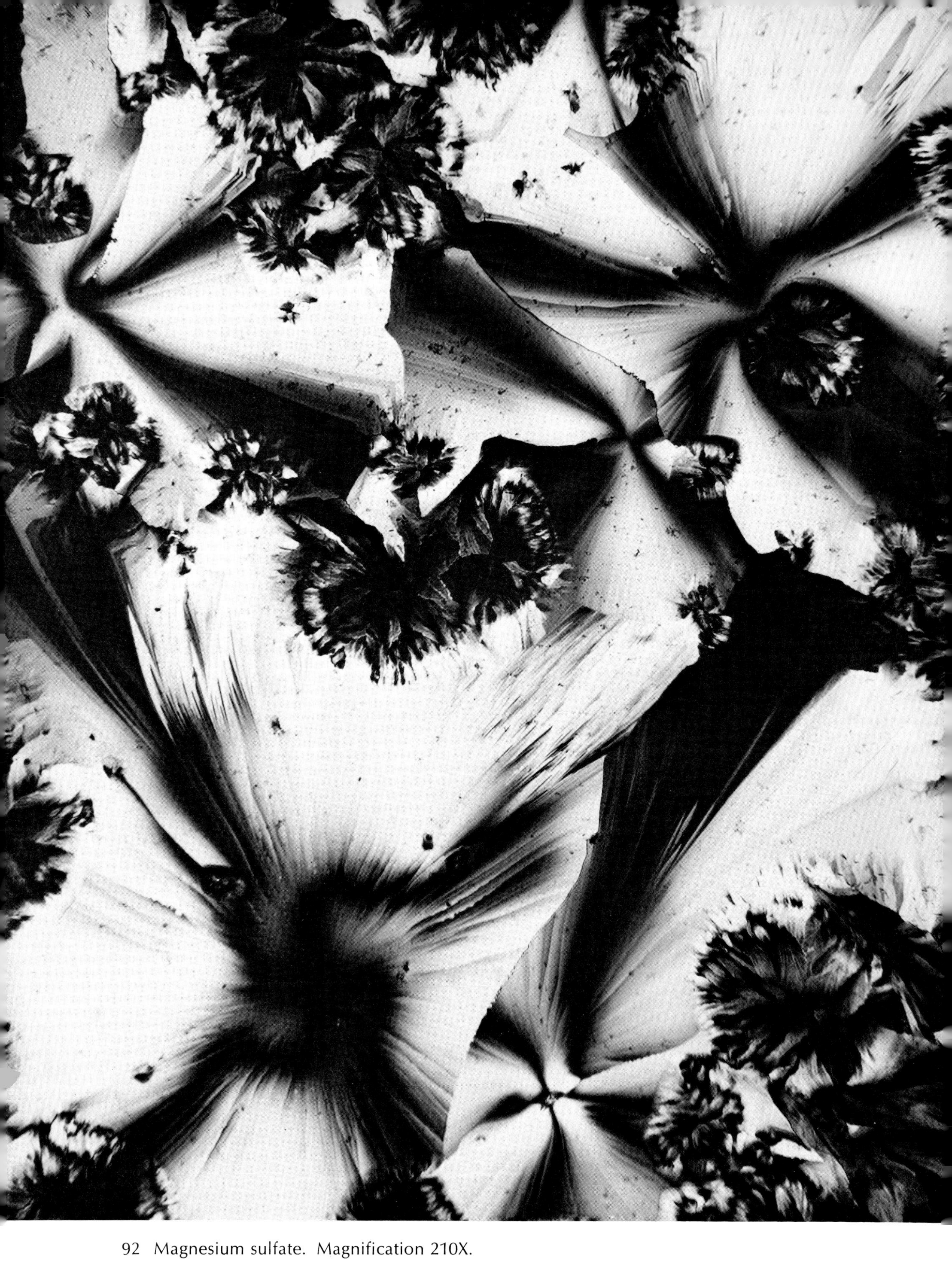

92 Magnesium sulfate. Magnification 210X.

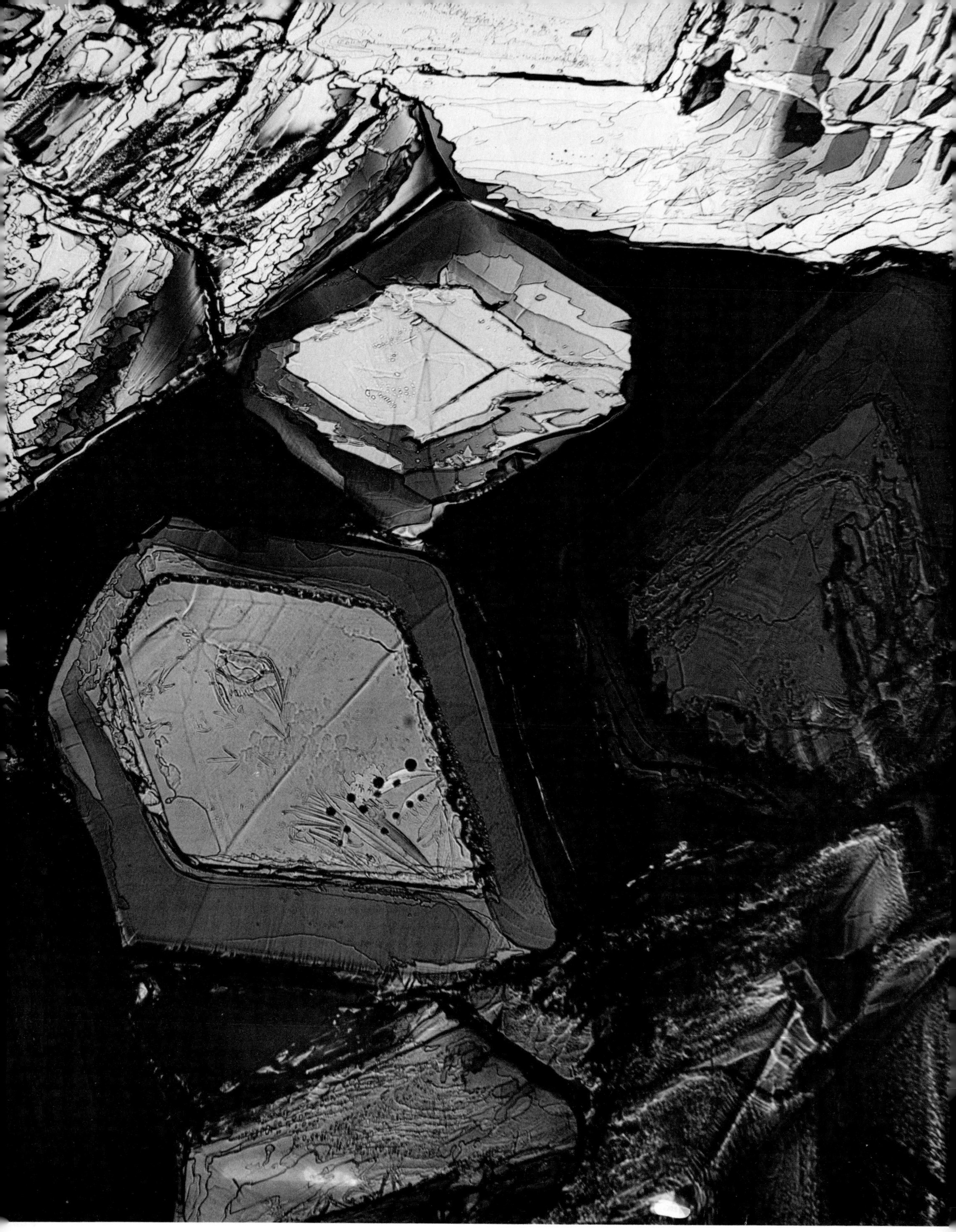

93 Beta-alanine. Magnification 225X.

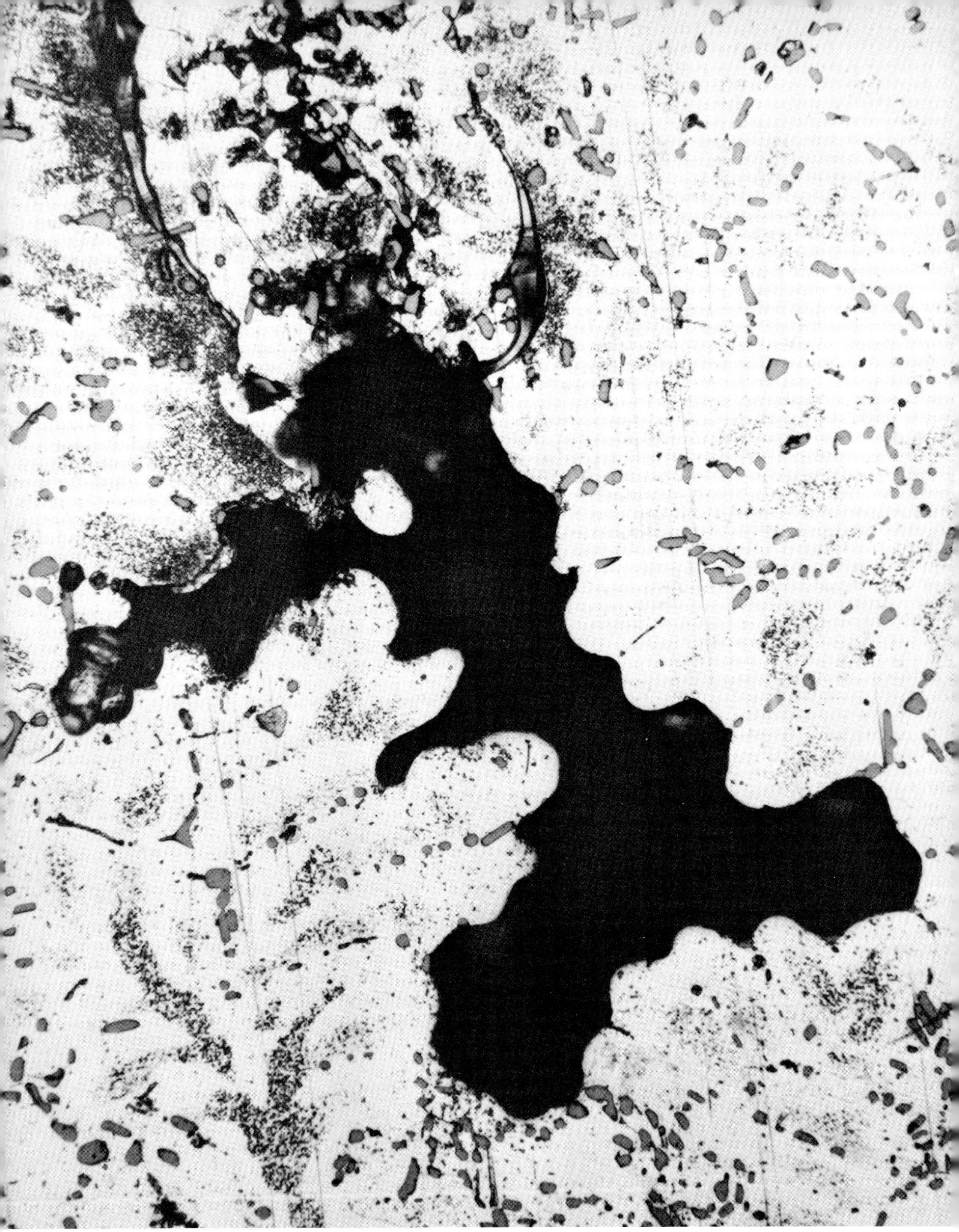

94 Cast-aluminum alloy. Magnification 875X.